D0053918

Tourism and Development in the Developing World

Tourism is widely considered as an effective contributor to socio-economic development, particularly in less developed countries. However, despite the almost universal adoption of tourism as a developmental option, the extent to which economic and social development inevitably follows the introduction and promotion of a tourism sector remains the subject of intense debate. The purpose of this book, therefore, is to provide an introduction to the tourism–development process. Focusing specifically on the less developed world and drawing on contemporary case studies, it questions many assumptions about the role of tourism in development and, in particular, highlights the dilemmas faced by destinations seeking to achieve development through tourism.

An introductory chapter establishes a foundation for the book, exploring the meaning and objectives of development, reviewing theoretical perspectives on the developmental process, and assessing the reasons why less developed countries are attracted to tourism as a development option. The concept of sustainable development, as the most widely adopted contemporary model of development, is then introduced and its links with tourism critically appraised. Subsequent chapters then address key issues related to tourism and development, including: the relationship and interaction between the forces of tourism and globalization; the tourism development process; the relationship between tourism and the communities within which it is developed; the management implications of trends and changes in the demand for and consumption of tourism; and, an analysis of the consequences of tourism development for destination environments, economies and societies. Finally, the issues raised throughout the book are drawn together in a concluding chapter that reviews the tourism and development 'dilemma'.

Combining an overview of essential concepts, theories and knowledge related to tourism and development with an analysis of contemporary issues and debates, the book is a valuable resource for those investigating tourism issues in developing countries. It is also useful for students studying related subjects, including development studies, geography, international relations, politics, sociology, and area studies.

David J. Telfer is Associate Professor in the Department of Tourism and Environment at Brock University, Canada.

Richard Sharpley is Professor of Tourism and Head of Department of Tourism and Recreation at the University of Lincoln.

Routledge Perspectives on Development

Series Editor: Professor Tony Binns, *University of Otago*

The *Perspectives on Development* series will provide an invaluable, up-to-date and refreshing approach to key development issues for academics and students working in the field of development, in disciplines such as anthropology, economics, geography, international relations, politics and sociology. The series will also be of particular interest to those working in interdisciplinary fields, such as area studies (African, Asian and Latin American Studies), development studies, rural and urban studies, travel and tourism.

If you would like to submit a book proposal for the series, please contact Tony Binns on j.a.binns@geography.otago.ac.nz

Published:

Third World Cities, 2nd edition
David W. Drakakis-Smith

Rural–Urban Interactions in the Developing World
Kenneth Lynch

Children, Youth and Development
Nicola Ansell

Theories and Practices of Development
Katie Willis

An Introduction to Sustainable Development, 3rd edition
Jennifer A. Elliott

Environmental Management and Development
Chris Barrow

Gender and Development
Janet Henshall Momsen

Tourism and Development in the Developing World
David J. Telfer and Richard Sharpley

Forthcoming:

Cities and Development
Jo Beall

Health and Development
Hazel Barrett

Local Knowledge, Environment and Development
Tony Binns, Christo Fabricius and Etienne Nel

Participation and Development
Andrea Cornwall

Disaster and Development
Andrew Collins

Postcolonialism and Development
Cheryl McEwan

Conflict and Development
Andrew Williams and Roger MacGinty

Africa: Diversity and Development
Tony Binns and Alan Dixon

Economics and Development Studies
Michael Tribe, Frederick Nixon and Andrew Sumner

Water Resources and Development
Clive Agnew and Philip Woodhouse

Global Finance and Development
David Hudson

Population and Development
W.T.S. Gould

Southeast Asian Development
Andrew McGregor

Tourism and Development in the Developing World

David J. Telfer
and
Richard Sharpley

LONDON AND NEW YORK

First published 2008
by Routledge
2 Park Square, Milton Park, Abingdon, Oxon OX14 4RN

Simultaneously published in the USA and Canada
by Routledge
270 Madison Ave, New York, NY 10016

Routledge is an imprint of the Taylor & Francis Group, an informa business

Typeset in Times New Roman by
RefineCatch Limited, Bungay, Suffolk
Printed and bound in Great Britain by
TJ International Ltd, Padstow, Cornwall, UK

British Library Cataloguing in Publication Data
A catalogue record for this book is available from the British Library

Library of Congress Cataloging in Publication Data
Sharpley, Richard, 1956–
Tourism and development in the developing world / by Richard Sharpley and David J. Telfer.
p. cm.
Includes bibliographical references and index.
ISBN 978–0–415–37144–5 (hardcover)—ISBN 978–0–415–37151–3 (softcover)—ISBN 978–0–203–93804–1 (e-book) 1. Tourism—Developing countries. 2. Tourism—Economic aspects—Developing countries. 3. Sustainable development—Developing countries. I. Telfer, David J. II. Title.
G155.A1S476 2008
338.4′791091724—dc22
2007025681

ISBN13: 978–0–415–37144–5 (hbk)
ISBN13: 978–0–415–37151–3 (pbk)
ISBN13: 978–0–203–93804–1 (ebk)

To Olivia, Rosie, Kyoko and Sakura

Contents

List of plates viii
List of figures x
List of tables xi
List of boxes xii
Preface xiii
Acknowledgements xv

1 Introduction: tourism in developing countries 1

2 Tourism and sustainable development 30

3 Globalization and tourism 57

4 The tourism planning and development process 80

5 Community response to tourism 115

6 The consumption of tourism 146

7 Assessing the impacts of tourism 174

8 Conclusion: the tourism development dilemma 205

References 233
Index 255

Plates

1.1	Cuba, Varadero: Hotel construction	18
1.2	Tunisia, near Monastir: Hotel construction	18
1.3	Indonesia, Yogyakarta: A woman whose family owns and operates a small hotel returning from a traditional market	20
1.4	China: Tourists on Great Wall of China	21
2.1	South Africa, Dikhololo Resort near Pretoria: Tourists preparing to go on a game-watching outing	41
2.2	Russia, St Petersburg: Code of conduct for tourists visiting the Peter and Paul Fortress	52
3.1	Nassau, Bahamas: Multiple cruise ships in port	68
3.2	Indonesia, Lombok: Sign indicating where a future Holiday Inn will be built	68
4.1	Cuba, Varadero: Beach resort	100
4.2	South Africa, Sun City Resort: Tourists swimming at the man-made beach at the resort	100
4.3	Tunisia, Monastir: Luxury beach resort, Amir Palace Hotel	101
4.4	Indonesia, Lombok: Fisherman-turned-supplier purchases fish in a local fish market for an international hotel	103
4.5	Indonesia, Lombok: Small local fruit and vegetable supplier makes a delivery to the Sheraton Hotel in Sengiggi Beach	104
4.6	Thailand, Bangkok: Tourists visiting the Grand Palace Complex	107
4.7	Tunisia, El Jem: Roman Colosseum	109
5.1	Argentina, Estancia Santa Susana, near Buenos Aires: Tour guide at a historic ranch presents traditional implements	118

5.2	Indonesia, village of Bangunkerto: The site of community-based agritourism project	125
5.3	Cuba, La Moka Ecolodge	127
5.4	Cuba, Las Terrazas: Located next to La Moka Ecolodge, the community of Las Terrazas is home to some of the people who work in the resort	127
5.5	South Africa, Township near Pretoria: Tourists visit a Township	140
5.6	Indonesia, Lombok: Young women present traditional Sasak weaving	143
6.1	Tunisia, near Matmata: These camels are used to provide rides to tourists	158
6.2	Cuba, Havana: Horse and buggy rides for tourists can be contrasted with the local citizens' form of transportation	160
6.3	Nassau, Bahamas: Tourists walk through the Prince George Wharf area	161
6.4	South Africa, near Pretoria: Shops selling souvenirs to tourists	171
7.1	Indonesia, Bali: Local entrepreneurs are making dyed textiles	184
7.2	Argentina, Iguazu Falls: A World Heritage Site	186
7.3	Argentina, Estancia Santa Susana, near Buenos Aires: Cultural performance at historic ranch	194
7.4	Indonesia, Bali: Traditional cultural ceremonies	197
8.1	Indonesia, Lombok: Local village very close to main tourist resort area	212
8.2	Indonesia, Kuta Beach Bali: Traditional Balinese ceremony on the beach	227

Figures

1.1	Influences on tourism's development	3
4.1	The tourism development process	83
6.1	The tourism demand process	149
6.2	Cohen's typology of tourists (1972)	151
6.3	A code of ethics for tourists	166
7.1	Impacts of tourism: a framework for analysis	177
7.2	Tourist experience of destination environment	178
7.3	Factors that influence the economic impacts of tourism	181
7.4	The tourism multiplier process	182
7.5	A model for assessing tourism's physical impacts	188
7.6	Community responses to tourism impacts	199
7.7	Baseline issues and indicators for sustainable tourism development	202
8.1	The tourism development dilemma framework	226

Tables

1.1	Worldwide export earnings, 2002	2
1.2	Per capita GNI country classifications	8
1.3	Standard of living indicators, by total numbers (millions), 2000	10
1.4	The evolution of development theory	12
1.5	International tourism arrivals and receipts growth rates, 1950–2000	17
1.6	International tourist arrivals and receipts, 1950–2002	22
1.7	World's top ten international tourism destinations, 2003	22
1.8	World's top ten international tourism earners, 2003	23
1.9	World's top ten international tourism generators by expenditure	23
1.10	Percentage share of international tourist arrivals by region, 1960–2003	23
2.1	Sustainable development: principles and objectives	36
2.2	Characteristics of mass vs alternative tourism	39
2.3	Sustainable tourism development: a summary of principles	43
2.4	Agenda for sustainable tourism	51
3.1	Members of Star Alliance, Oneworld and Sky Team, 2006	66
3.2	Locations of Hard Rock Cafés in developing and transitional countries	69
4.1	Examples of tourism plans in Indonesia	94
5.1	Concerns over tourism interaction in developing countries and implications for communities	119
7.1	Travel and tourism economy as percentage of total GDP	185
8.1	Spheres of environmental activity and questions for sustainable tourism development	214

Boxes

1.1	The UN Millennium Project	7
1.2	Tourism and development in Kyrgyzstan	16
2.1	Tourism and sustainable development in Bhutan	37
2.2	The Damaraland Camp, Namibia	40
2.3	Tourism and development in The Gambia	46
3.1	Globalization and Cancún, Mexico	65
3.2	Globalization and the cruise ship industry	67
4.1	Fonatur and tourism development in Mexico	87
4.2	The UN World Tourism Organization and tourism planning consultants	90
5.1	Community-based ecotourism in Cuba	126
5.2	Pro-poor tourism in South Africa	139
6.1	All-inclusive resorts – the case of Sandals, Jamaica	153
6.2	Tourism development in Dubai	162
6.3	Ecotourists in Belize	165
7.1	Impacts of trekking in the Nepalese Himalaya	190
7.2	The commoditization of dance masks, Sri Lanka	196
8.1	Tourism development and human rights	208

Preface

Tourism is increasingly viewed as an attractive development option for many parts of the developing world. In some developing nations, it may in fact be the only viable means of stimulating development. However, as developing countries opt into this industry they face what is referred to in this book as a tourism–development dilemma.

Developing nations are seeking the potential benefits of tourism, such as increased income, foreign exchange, employment and economic diversification; nevertheless, these developmental benefits may in fact fail to materialize. In entering this global competitive industry, developing countries may find tourism benefits only the local élite or multinational corporations, or is achieved at significant economic, social or environmental costs. The challenge in this dilemma is then accepting or managing the negative outcomes of the tourism–development process for the potential long-term benefits offered by tourism.

The purpose of this book is to explore the nature of the tourism–development dilemma by investigating the challenges and opportunities facing developing countries pursuing tourism as a development option. The book begins with an examination of the nature of developing countries and why they are attracted to such a volatile industry as a preferred development tool. It is important to consider to what extent tourism can contribute to overall development broadly defined, and so Chapter 1 also examines the evolution of developmental thought whereby development is no longer tied solely to economic criteria. The second chapter examines the nature of sustainable development and its relationship to tourism, which has become a much-contested concept focusing not only on the physical environment but also on the economic, social and cultural environment. This chapter sets the stage for the remainder of the text by raising key issues, including the influence of globalization on tourism (Chapter 3), the tourism planning and development process (Chapter 4), community responses to tourism (Chapter 5), consumption of tourism (Chapter 6), and an analysis of tourism impacts (Chapter 7). The concluding chapter draws together the main issues in the book, presenting a tourism–development dilemma framework that illustrates the complexity of often-interconnected forces at work in using tourism as a development tool. While it is argued that there is a development imperative and a sustainable development imperative, it is important to recognize the challenges of implementing the ideals of sustainability in the context of the realities in the tourism industry in developing countries.

The focus of the book is to present an introductory-level text that explores the relationship between tourism and development, and it is designed in part to be a successor to John Lea's *Tourism and Development in the Third World* (1988), originally published in the Routledge Introductions to Development series.

Acknowledgements

The authors would like to thank Andrew Mould and his colleagues at Routledge for all their patience and assistance with this project. We would also like to thank Sandra Notar for help with proofreading; Hui Di Wang for Plate 1.4; and Tom and Hazel Telfer for Plates 5.1, 7.2 and 7.3. Finally we would also like to thank, as always, Julia Sharpley and Atsuko Hashimoto for their support during the writing of this book.

The cover photo (beach vendor in Cuba) is by Richard Sharpley and other photos are by David J. Telfer unless otherwise indicated.

1 Introduction: tourism in developing countries

Learning objectives

When you have finished reading this chapter, you should be able to:

- **Appreciate the characteristics of underdevelopment in developing countries;**
- **Understand why tourism is selected as a development option for developing countries;**
- **Identify global tourism market shares and the changing nature of tourism;**
- **Be familiar with the different approaches to tourism and development.**

Over the past half century, tourism has evolved into one of the world's most powerful, yet controversial, socio-economic forces. As ever greater numbers of people have achieved the ability, means and freedom to travel, not only has tourism become increasingly democratized (Urry 2001) but also both the scale and scope of tourism have grown inexorably. In 1950, for example, just over 25 million international tourist arrivals were recorded worldwide; by 2004, the UK alone was attracting that number of overseas visitors annually while, according to the World Tourism Organization (WTO 2005a), total international arrivals reached a record 760 million. Moreover, if domestic tourism (that is, people visiting destinations within their own country) is included, the total global volume of tourist trips is estimated to be between six and ten times higher than that figure. Little wonder, then, that tourism has been described as the 'largest peaceful movement of people across cultural boundaries in the history of the world' (Lett 1989: 265).

As participation in tourism has grown, so too has the number of countries that play host to tourists. Although approximately half of all international arrivals are still received by just ten (principally developed) nations, many new destinations have claimed a place on the international tourism map while numerous more distant, exotic places have, in recent years, enjoyed a rapid increase in tourism. Indeed, throughout the past decade the East Asia Pacific and Middle East regions have witnessed the highest and most sustained growth in arrivals globally while, in particular, a number of least developed countries (LDCs), including Cambodia, Myanmar, Samoa and Tanzania, have experienced higher than world average growth in tourism (UNCTAD 2001). Such is the global scale of tourism that the WTO currently publishes annual tourism statistics for around 200 countries.

Reflecting this dramatic growth in scale and scope, tourism's global economic contribution has also become increasingly significant. International tourism alone generated over US$523 billion in 2003 and, if current forecasts prove to be correct, this figure could rise to US$2 trillion by 2020 (WTO 1998). By the end of the twentieth century, tourism also represented the world's most valuable export category, although more recently it has fallen back to fourth place, currently accounting for 7 per cent of worldwide exports of goods and services (Table 1.1). According to the World Travel and Tourism Council, if domestic tourism is added, tourism as a whole generates over US$4 trillion annually, accounting for approximately 10 per cent of global GDP and employment.

Table 1.1 Worldwide export earnings, 2002

		US$ billion	**%**
	Total exports of goods and services	7,903	100.0
1	Chemicals	660	8.4
2	Automotive products	621	7.6
3	Fuels	615	7.3
	– International tourism receipts	474	6.0
	– International fare receipts	104	1.3
4	*Total international tourism*	*578*	*7.3*
5	Computers/office equipment	491	6.2
6	Food	468	5.9
7	Textiles/clothing	353	4.5
8	Telecommunications equipment	347	4.4

Source: Adapted from WTO (www.world-tourism.org/facts/trends/economy.htm).

Given this remarkable growth and economic significance, it is not surprising that tourism has long been considered an effective means of achieving economic and social development in destination areas. Indeed, the most common justification for the promotion of tourism is its potential contribution to development, particularly in the context of developing countries. That is, although it is an important economic sector and often a vehicle of both rural and urban economic regeneration in many industrialized nations, it is within the developing world that attention is most frequently focused on tourism as a developmental catalyst. In many such countries, not only has tourism become an integral element of national development strategies (Jenkins 1991) – though often, given a lack of viable alternatives, an option of 'last resort' (Lea 1988) – but also an increasingly important sector of the economy, providing a vital source of employment, income and foreign exchange, as well as a potential means of redistributing wealth from the richer nations of the world. For example, as the 2001 UN Conference on Trade and Development noted, 'tourism development appears to be one of the most valuable avenues for reducing the marginalization of LDCs from the global economy' (UNCTAD 2001: 1).

Importantly, though, the introduction of tourism does not inevitably set a nation on the path to development. In other words, many developing countries are, at first sight, benefiting from an increase in tourist arrivals and consequential foreign exchange earnings. However, the unique characteristics of tourism as a social and economic activity, and the complex relationships between the various elements of the international tourism system and transformations in the global political economy of which it is a part, all serve to reduce its potential developmental contribution. Not only is tourism highly susceptible to external forces and events, such as political upheaval (e.g. Fiji, which has suffered a number of military coups, the most recent in December 2006), natural disasters (e.g. the Indian Ocean tsunami in December 2004), terrorist attacks (e.g. Bali, Indonesia in 2002 and 2005, or the decline in travel following the September 2001 attacks in the USA) or health scares (e.g. SARS in 2003), but many countries have become increasingly dependent on tourism as an economic sector which remains dominated by wealthier, industrialized nations (Reid 2003). Moreover, the political, economic and social structures within developing countries frequently restrict the extent to which the benefits of tourism development are realized. The factors that influence tourism's potential developmental contribution are summarized in Figure 1.1.

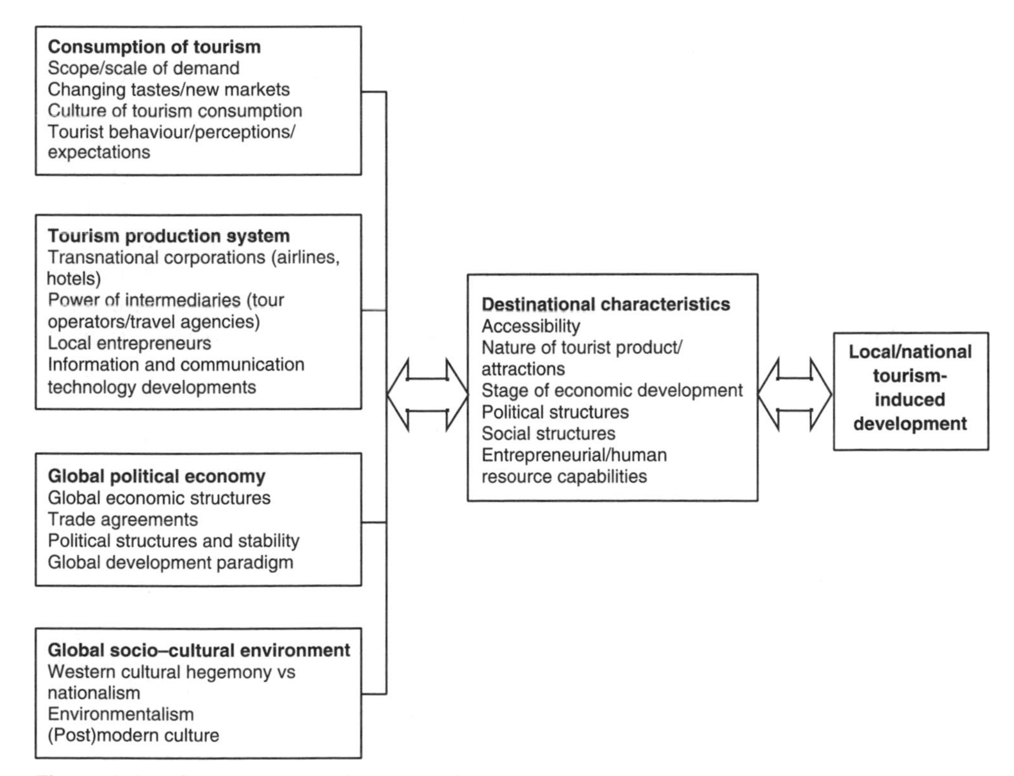

Figure 1.1 ***Influences on tourism's development.***

Many of these issues will be addressed throughout this book. However, the fundamental point is that there exists what may be referred to as a tourism–development dilemma. That is, tourism undoubtedly represents a potentially attractive (and frequently, the only viable) means of stimulating social and economic development in destination areas and nations; yet often that development fails to materialize, benefits only local élites, or is achieved at significant economic, social or environmental cost to local communities. The dilemma for many developing countries, therefore, lies in the challenge of accepting or managing such negative consequences for the potential longer-term benefits offered by tourism development.

The purpose of this book is to explore the challenges and opportunities facing developing countries that pursue tourism as a development option. In so doing, it will critically appraise contemporary perspectives on tourism and development and, in particular, sustainable tourism development as the dominant tourism development paradigm. However, the first task is to consider the concepts of underdevelopment/development and the relevance of tourism as a development option. It is with this issue that the remainder of this introductory chapter is concerned.

Focus and definitions

As noted above, this book is primarily concerned with tourism in developing countries. The term 'developing country' is, of course, subject to wide interpretation and is often used interchangeably with other terminology, such as 'Third World' or 'less developed country' or, more generally, 'The South'. However, it usefully contrasts a country or group of countries (the 'developing world') with those that are 'developed', although, similarly, there is no established convention for defining a nation as 'developed'. Nevertheless, the developed countries of the world – those that are technologically and economically advanced, enjoy a relatively high standard of living and have modern social and political structures and institutions – are generally considered to include Japan, Australia and New Zealand in Oceania, Canada and the USA in North America, and the countries that formerly comprised Western Europe. Some commentators also include Israel, Singapore, Hong Kong and South Korea as developed countries.

Of course, categorizing the countries of the world as either 'developed' or 'developing' oversimplifies a complex global political economy. Not only does the developing world include countries that vary enormously in terms of economic and social development, with some, such as Brazil, the South-east Asian 'tiger' economies and, of course, China and India, assuming an ever-increasingly important position in the global economy, but new trade or political alignments cut across the developed–developing dichotomy. For example, the G-20, or Group of Twenty, established in 1999, promotes dialogue between industrialized and those emerging market countries not considered to be adequately included in global economic discussion and governance (see www.g20.org). The group comprises Argentina, Australia, Brazil, Canada, China, France, Germany, India, Indonesia, Italy, Japan, Mexico, Russia, Saudi Arabia, South Africa, South Korea, Turkey, the

UK and the USA; countries that account for 90 per cent of global GDP and 80 per cent of world trade.

Nevertheless, for the purposes of this book, the term 'developing country' embraces all nation states that are not generally recognized as being developed, including the transitional economies of the former 'Second World' and contemporary, centrally planned economies. Although it covers an enormous diversity of countries which may demand subcategorization, this focus reflects Britton's (1982) metropolitan/periphery political–economic model which, arguably, still defines the structure of international tourism. Certainly, recent figures reflect the continuing domination of the developed world in terms of both international arrivals and receipts: in 2002, the developed countries collectively accounted for approximately 54 per cent of arrivals and 61 per cent of receipts, the latter showing little change since 1997 when the industrialized countries attracted 63 per cent of all international tourism receipts.

The terms 'tourism' and 'development' also require definition. Regarding tourism, most introductory texts consider the issue in some depth (see e.g. Sharpley 2002; Cooper *et al.* 2005), while generally, many definitions have been proposed. However, these may be classified under two principal headings:

- ***Technical definitions***. These attempt to identify different categories of tourist for statistical or legislative purposes. Various parameters have been established to define a tourist, such as minimum (one day) and maximum (one year) length of stay, minimum distance travelled from home (160km) and purpose, such as 'holiday' or 'business' (WTO/UNSTAT 1994), though a useful overall definition has been proposed by the UK Tourism Society:

  > Tourism is the temporary short-term movement of people to destinations outside the places where they normally live and work, and their activities during their stay at these places; it includes movement for all purposes, as well as day visits or excursions.

- ***Conceptual definitions***. These attempt to convey the meaning or function of tourism as a particular social institution (see Burns and Holden 1995; Sharpley 2003). Typically, these emphasize the nature of tourism as a leisure activity that contrasts with normal, everyday life (perhaps the most commonly held perception of what tourism is) and provide a basis for assessing tourist behaviour and attitudes.

Development is a more complex concept and one that 'seems to defy definition' (Cowen and Shenton 1996: 3). Moreover, according to the post-development school, it is also a global concept that, over the last half century, has failed in its objectives and should therefore be abandoned (Rahnema and Bawtree 1997). Nevertheless, development remains a term in common use, referring to both a process societies undergo and the goal or outcome of that process – that is, the development process in a society may result in its achieving the state or condition of development. It is also a term, albeit one usually considered in the context of developing countries, that relates to every nation in the world. In other words, a society that is 'developed' does not cease to change or progress; the nature of that change may, however, be different to change in less developed societies.

Traditionally, development was measured in economic terms, typically GNP or per capita GDP. Indeed, during the 1950s and 1960s, development and economic growth were considered to be synonymous (Mabogunje 1980). However, as Seers (1969) argued, this revealed nothing about improvements (or lack of) in the distribution of wealth, the reduction of poverty, employment and other factors such as education, housing, healthcare and so on. Thus, development has become a much broader concept embracing at least five dimensions (see Goulet 1992):

- *An economic component* – wealth creation and equitable access to resources;
- *A social component* – improvements in health, housing, education and employment;
- *A political dimension* – assertion of human rights, appropriate political systems;
- *A cultural dimension* – protection or affirmation of cultural identity and self-esteem;
- *The full-life paradigm* – preservation and strengthening of a society's symbols, beliefs and meaning systems.

To these should be added perhaps an ecological component, not only reflecting the emergence of environmental sustainability as a fundamental parameter of contemporary approaches to development but also, as discussed in Chapter 2, the basis of the concept of sustainable tourism development. Collectively, these dimensions are broadly reflected in the UN Millennium Project's goals and targets (see Box 1.1), while today, the most widely accepted measure of development is the annual UNDP Human Development Index (HDI) which ranks countries according to a variety of economic and social indicators (see also Dasgupta and Weale 1992).

Development, then, is a complex, multidimensional concept that may be defined as a continuous and positive change in the economic, social, political and cultural dimensions of the human condition, guided by the principle of freedom of choice and limited by the environment's capacity to sustain such change. The question to be addressed now, however, is: What are the particular characteristics of any society or country that define it as underdeveloped?

Underdevelopment and development

Many of the problems facing developing countries, such as poverty, inequality, poor healthcare and a lack of educational opportunities, are widely recognized; they are also reflected in the goals of international development programmes such as the UN Millennium Project. Moreover, development is arguably regaining a dominant position within international politics, embodied in the international fight against poverty and the contemporary *Make Poverty History* campaign. However, the specific characteristics of underdevelopment are less clear; that is, many of the

Box 1.1

The UN Millennium Project

The United Nation's Millennium Project is an independent advisory body that draws on research undertaken by ten Task Forces, collectively comprising more than 260 development experts, to advise the UN on appropriate strategies to achieve an internationally agreed set of global development targets, including reducing poverty, hunger, disease and environmental degradation, by 2015. If these goals are met, it is claimed that half a billion people will be lifted out of poverty and a further 250 million will no longer suffer from hunger. The Project has eight Millennium Development goals (MDGs), each of which has specific targets to be met by 2015.

Goal 1: Eradicate extreme poverty and hunger
Target 1: reduce by half the proportion of people whose income is less than $1 a day
Target 2: reduce by half the proportion of people who suffer from hunger

Goal 2: Achieve universal primary education
Target 3: ensure that children everywhere are able to complete full primary schooling

Goal 3: Promote gender equality and empower women
Target 4: eliminate gender disparity in all levels of education

Goal 4: Reduce child mortality
Target 5: reduce the under-5 mortality rate by two-thirds

Goal 5: Improve maternal health
Target 6: reduce the maternal mortality rate by three-quarters

Goal 6: Combat HIV/AIDS, malaria and other diseases
Target 7: halt/reverse the spread of HIV/AIDS
Target 8: halt/reverse the incidence of malaria and other serious diseases

Goal 7: Ensure environmental sustainability
Target 9: integrate the principles of sustainable development into national development policies
Target 10: halve the proportion of people without access to basic sanitation and drinking water
Target 11: achieve a significant improvement in the lives of 100 million slum dwellers

Goal 8: Develop a global partnership for development
Target 12: develop an open, non-discriminatory trading and financial system
Target 13: address the special needs of least developed countries
Target 14: address the special needs of landlocked developing countries and small island developing states
Target 15: deal comprehensively with the developing countries' debt problems

Source: Adapted from www.unmillenniumproject.org.

problems facing developing countries are the outcome, not the cause, of under-development. Consequently, it is also unclear to what extent particular developmental vehicles, such as tourism, are effective as means of addressing these problem and challenges.

It is also important to point out that there is an enormous diversity of countries that comprise the developing world as defined for the purposes of this book. Geographical, political, historical, economic and socio-cultural characteristics and structures all influence a country's level or rate of development (Todaro 2000), as well as its tourism development potential. However, as noted above, developing countries are typically classified according to national and/or per capita income, non-economic development indicators, such as life expectancy or literacy, or environmental factors, or a combination of the two. The World Bank, for example, classifies all countries according to per capita gross national income (see Table 1.2), accepting that level of income does not necessarily reflect development status. Consequently, a number of the 56 countries classified as high income are not generally considered to be among the group of developed nations.

Table 1.2 Per capita GNI country classifications

	Low income economies	Lower-middle income economies	Upper-middle income economies	High income economies
Per capita GNI	$735 or less	$736–$2,935	$2,936–$9,075	$9,076 or above
Number of countries in group	64	54	34	56

Source: World Bank (2005).

The term least developed countries (LDCs) is also used increasingly to distinguish the world's poorest nations from a total of approximately 160 developing countries. To be added to the list of LDCs (which, in the review of 2003, comprised a total of 50 states) a country must have a per capita income below $750, as well as satisfying complex 'economic vulnerability' and 'human resource weakness' criteria. Many of these countries are referred to by de Rivero (2001) as NNEs, or 'non-viable national economies', suggesting that they cannot be regarded as 'developing' countries in any sense of the word. Interestingly, a number of LDCs have either established or nascent tourism sectors which, though small by international standards, are significant in terms of the local economy – the Maldives, for example, attracted 485,000 visitors in 2002, with tourism representing about 75 per cent of the islands' GDP. Similarly, in the West African countries of The Gambia and Senegal, the tourism sector is small in terms of arrivals (75,000 and 420,000 respectively), yet in both countries it contributes some 50 per cent of total exports. In these cases, however, tourism may be considered to be economic survival as opposed to development strategy.

The characteristics of underdevelopment

Developing countries typically share a number of features that characterize the conditions of underdevelopment.

- *Economic dependence on a large, traditional agricultural sector and the export of primary products.* Most developing countries' economies are dependent on agricultural production and exports for employment, income and foreign exchange earnings. Conversely, the industrial/manufacturing sector may be small and technologically deficient. Typically, over 60 per cent of the workforce is employed in agriculture in developing countries compared with less than 5 per cent in developed countries. At the same time, low productivity and international price support mechanisms limit their ability to compete in global markets for primary products.
- *Low standards of living.* A variety of factors contribute to low standards of living. Reference has already been made to low income levels (per capita income), although it is important to point out that average income gives no indication of income inequality within countries. It has been observed that few developing countries 'enjoy the luxury of having less than 20 per cent of their population below the poverty line' (de Rivero 2001: 64), and in some countries the contrast is stark. In India, over 50 per cent of the population live in poverty (on an income of less than $1 a day) while in China, the world's fastest growing economy, the figure is around 30 per cent. Moreover, in many developing countries, the average level of income is falling: 46 countries are poorer now than they were in 1990 (UNDP 2004), while in some, a greater proportion of the population live below the national poverty line. In Hungary, for example, the number of people living in poverty increased by 2.8 per cent between 1993 and 1997.

  In addition to low income, other indicators of living standards include health, child mortality, education/literacy levels, access to clean water and so on. Table 1.3 provides details of selected indicators for regions of the developing world.
- *High population growth and high unemployment/underemployment.* Over 80 per cent of the world's population live in developing countries, a proportion that will continue to grow given higher birth rates on average (around 2 per cent annually) than in developed countries (about 0.5 per cent). In the period between 1995 and 2025, the population of many developing countries will double. Consequently, under- and unemployment in developing countries, averaging between 8 and 15 per cent of the workforce (though often double this figure among the 15 to 24 age group) will increase significantly.
- *Economic fragility.* The economies of many developing countries are weak, characterized by low financial reserves, severe balance of payment deficits and high levels of international debt. Limited natural resources and industrial production necessitate high levels of imports to meet basic needs, yet exports typically cover around only two-thirds of developing countries' import bills. The subsequent levels of international debt and interest

Table 1.3 Standard of living indicators by total numbers (millions), 2000

	Living on less than $1/day	Total under-nourished	Children under 5 dying each year	Primary-age children not in school	Without access to improved water supply
Sub-Saharan Africa	323	185	5	44	273
Arab States	8	34	1	7	42
East Asia and Pacific	261	212	1	14	453
South Asia	432	312	4	32	225
Latin America/ Caribbean	56	53	0	2	72
Central/Eastern Europe and CIS	21	33	0	3	29
Total	1,100	831	11	104	1,197

Source: Adapted from UNDP (2004).

payments have resulted in many developing countries becoming ensnared in a debt trap; hence the frequent calls for debts to be written off by Western creditors.

- *Limited or unstable socio-political structures.* While underdevelopment is frequently claimed to result from inequalities in the global distribution of economic and political power (with international tourism widely seen as a manifestation of such inequality), the political and social structures within developing countries may also determine the extent to which development occurs. Although the past quarter century has witnessed the dramatic spread of democratization (Potter 2000) – though not necessarily with a corresponding increase in development – the distribution of power in developing countries tends to favour a small, powerful élite. In many cases the position of the élite may be strengthened and legitimized by the apparent legitimate democratic process. This may determine the nature of development both generally and in the specific context of tourism, whether at national (Din 1982) or local level (Southgate 2006).

Inevitably, these characteristics of underdevelopment are not equally evident in all developing countries, while other indicators, such as gender-related issues (Momsen 2004), the ability to exercise human rights, or safety and security must also be included as measures of development. Moreover, many developed nations also have 'less developed' regions and face a number of developmental challenges, whether environmental, social (crime, inequality, education, health) or economic (poverty, unemployment). Nevertheless, within the developing world tourism is increasingly viewed as a means of addressing underdevelopment which, by implication, suggests that tourism may also impact positively on some or all of these specific challenges. The extent to which this occurs in practice is, of course, the subject of this book.

Development paradigms

Having introduced the characteristics of underdevelopment, it is also important to review briefly how development theory (that is, a combination of the ideological *ends* of development and the strategic *means* of achieving them) has evolved over time. A full consideration of development theory can be found elsewhere, both within the development literature (e.g. Hettne 1995; Preston 1996; Todaro 2000; Desai and Potter 2002) and the tourism literature (e.g. Telfer 2002a, in press). Importantly, however, just as the meaning of development has changed over the past half century, so too have the dominant perspectives, or paradigms, on how development may be encouraged or achieved. To a great extent, these have been reflected in the evolution of tourism development theories in particular, although, as will be discussed later in this chapter, the extent to which a causal relationship exists between development theory and tourism development (both theory and practice) is not always clear.

It should be noted that a number of what may be considered as 'sub-theories' of development also exist. These are, in effect, specific development policies that are normally followed at national or regional levels as opposed to the more overarching or 'grand' theories of development. For example, import substitution policies were dominant in Latin America in the 1950s and 1960s while, more recently, state-led or 'statist' approaches to development have enjoyed a resurgence in many countries (Wade 2004; see also Clancy 1999). Here, however, we are concerned with broader development paradigms and their relationship to tourism development in particular.

In general, the 'story' of development theory is one of a shift from traditional, top-down economic growth-based models to a more broad-based approach with an emphasis on bottom-up planning, the supplying of basic human needs and a focus on sustainable development. Telfer (2002a) examined four identifiable development paradigms that have emerged chronologically since the end of the Second World War and their relationship to tourism. These development paradigms are summarized in Table 1.4. However, as will be discussed below, development theory is at a crossroads. The search for a new paradigm is underway and the shape that development thought will take in the coming years is open to debate (Rapley 2002). Telfer (in press) examined a range of emerging concepts in development studies and these are also included in Table 1.4. In examining Table 1.4 it is important to remember that emerging paradigms do not necessarily replace preceding ones; despite criticisms, elements of each paradigm remain relevant today. The time lines are only guides as to when the paradigms gained prominence.

- *Modernization*. Modernization theory is based on the concept that all societies follow an inevitable evolutionary path from traditional to modern, characterized by a transformation from agriculture to industry, from rural to urban and from traditional to modern (i.e. Western) values and social institutions (Harrison 1988). Progress along this evolutionary path is considered to be dependent upon economic growth as the basis of

Table 1.4 The evolution of development theory

Time guide	Development paradigm/process	Theoretical perspectives and concepts
1950s–1960s	***Modernization***	Stages of growth: pass through Western development stages Diffusion: growth impulses/trickle-down effect
1950s–1960s	***Dependency***	Neo-colonialism: underdevelopment caused by exploitation by developed countries Dualism: poverty functional to global economic growth Structuralism: domestic markets, state involvement, import substitution
mid 1970s–1980s	***Economic neo-liberalism***	Free market: free competitive markets/ privatization Structural adjustment: competitive exports/ market forces One world: new world financial systems, deregulation
1970s–early 1980s	***Alternative development***	Basic needs: focus on food, housing, education, health Grassroots: people-centred development Gender: women in development, gender relations/empowerment Sustainable development: environmental management
1990s, 2000 and beyond	***Beyond the Impasse: the search for a new paradigm?***	Post-development: rejection of the concept of 'development' State-led development: more active role for the state Civil society and social capital: work of voluntary organizations including NGOs, connect citizens and state Transnational social movements: e.g. environmentalists, indigenous peoples, feminists, peace activists, etc. Culture studies: different worldviews are accommodated Development and security: conflict and chaos with state disintegration

Source: Adapted from Telfer (2002a: 39) and Telfer (in press).

development and, according to Rostow (1967), only when a country has reached the 'take-off stage' (manifested in the emergence of one or more significant industries which induce growth in associated sectors) can it begin to modernize or develop. A variety of theories and strategies are embraced by the modernization paradigm, although the focus is usually on the introduction of a 'growth pole' (an industry or economic sector) from which 'growth impulses' diffuse throughout the region, thereby stimulating modernization. The development of a tourist destination may be considered as one such growth pole.

- *Dependency*. Dependency theory, sometimes referred to as underdevelopment theory, arose in the 1960s as a critique of the modernization paradigm. Essentially, it proposes that underdevelopment results not from the particular socio-economic characteristics of less developed countries (as suggested by modernization theory) but from the external and internal political, economic and institutional structures that keep them in a dependent position relative to developed countries. In other words, global political and economic relations are such that wealthier, more powerful Western nations are able to exploit weaker, peripheral nations (often mirroring earlier colonial ties), thereby limiting developmental opportunities within less developed countries. Thus, underdevelopment can be explained by an unequal international capitalistic system within which developing countries are unable to 'break out of a state of economic dependency and advance to an economic position beside the major capitalist industrial powers' (Palma 1995: 162). Various theoretical perspectives on dependency theory exist, although tourism as a global industry largely dominated by Western businesses and tourists has long been considered a manifestation of the paradigm (Britton 1982; Bastin 1984; Nash 1989).
- *Economic neoliberalism*. As a reaction or 'counter-revolution' to interventionist Keynesian economic policy (Brohman 1996a), economic liberalism, which became popular during the Reagan–Thatcher era of the 1980s, espoused the role of international trade in export-led economic development. Proposing that the problems facing developing countries arose from excessive state intervention, its supporters argued that the path to development lay in promoting market liberalization, the privatization of state enterprises and the overall reduction of state intervention. As a result, international loan programmes administered by the World Bank and IMF to promote development were conditional upon adjustments to economic structures and political policies in recipient nations, hence the term 'Structural Adjustment Lending Programme' (SALP) (Mosely and Toye 1988). Tourism development in many countries has benefited from international structural funding (Inskeep and Kallenberger 1992), although SALPs have since been widely discredited (Harrigan and Mosely 1991). With the more recent shift to poverty reduction, SALPs have largely been replaced by Poverty Reduction Strategy Papers; however, they too have been criticized.

- *Alternative development.* Signifying a departure from (or an alternative to) the preceding Western-centric, economic growth-based development paradigms, alternative development adopts a resource-based, bottom-up approach that focuses primarily on human and environmental concerns. Recognizing that development is a complex, multilayered process embracing not only economic growth but broader social, cultural, political and environmental factors, its fundamental tenet is that development should be endogenous. That is, it is a process that starts within, and is guided by the needs of, each society; it is not something that other societies should implement or impose. It also emphasizes the importance of satisfying basic needs and encouraging self-reliance (Galtung 1986), while environmental management is also a key element. Since the late 1980s, the alternative development paradigm has become more widely adopted as sustainable development, a concept that, although highly contested, continues to dominate global development policy. Not coincidentally, the 1980s concept of 'alternative tourism' (Smith and Eadington 1992) also provided the basis for what has become the dominant tourism development paradigm; namely sustainable tourism development. This is considered in more detail in Chapter 2.
- *Beyond the Impasse: the search for a new paradigm?* Development studies in the 1980s underwent what has become known as an impasse, since the current development theories could no longer explain all the difficulties that developing countries were encountering. Schuurman's book *Beyond the Impasse* (1996) outlined the main reasons for this impasse as:

  1. The gap between rich and poor countries had continued to widen;
  2. Developing countries were more concerned with short-term policies and were unable to implement long-term policies;
  3. Economic growth was having a major impact on the environment, and advocates for sustainable development called for reduced growth;
  4. Socialism was no longer seen as a legitimate or viable political means of solving developmental problems;
  5. With the rise in globalization, traditional development theories focused on the state did not coincide with a reduction in state power;
  6. The recognition that there is not a homogeneous Third World;
  7. With the rise in postmodernism in social science, grand narratives or theories did not apply.

What has emerged out of the impasse in development studies is a diversity of voices and approaches, a few of which are illustrated in Table 1.4. It is interesting to note, however, that some more recent ideas share similarities with concepts from the past. In more of a critique of development, those in the post-development camp have rejected the term 'development' itself, since it reinforces inequalities. Others have rejected the free market neoliberal approach and have called for a more state-led approach. Clancy (1999) has explored the Mexican state-led approach in tourism (see also Box 4.1). Others continue to suggest that civil society and the links to democracy, as well as social capital, have a key role in development as the state downsizes, stressing the importance of building trust in communities, which has been linked to the work of NGOs (Rapley 2002). Transnational social movements, such as the environmentalist movement, have strong links to the alternative development paradigm described above and are often also seen in opposition to, and at times actively protesting against, the free market neoliberal approach (Helleiner 2006). The final two concepts in Table 1.4 include culture studies emphasizing the importance of accommodating different worldviews and the links between development and security. The links with culture here are in part associated with breaking away from Eurocentric development thinking to accommodate different worldviews on development (Hettne 2002).

The final concept considered here is related to security, an issue that has recently come to the forefront globally. For tourism to operate successfully, a secure and stable environment is needed. With such a diversity of voices and approaches offered on development, what is the way forward? Hettne (2002: 11) states:

> The emerging approach can be described as transcendence: development studies as a precursor of a comprehensive and universally valid historical social science, devoted to the contextual study of different types of societies in different phases of development, struggling to improve their structural positions within the constraints of one world economy and one, albeit multilayered, world order. Furthermore, development theory needs to be reconstructed in terms of content as well.

Hettne (2002) has suggested that emerging approaches, which may help in building towards the future and a global social theory, consist of elements of international political economy, the links between peace, conflict and development, a new cultural study emphasizing the relevance of alternative thinking, a continuing concern for the excluded and an investigation into the nature of the world order, since the framework for development will no longer focus solely on the state.

As this brief chronology of development theory demonstrates, both the process and objectives of development have evolved over time, from relatively simplistic economic growth models to the more complex notion of sustainable development and the diversity of approaches more recently presented. As will be discussed shortly, approaches to tourism development have also evolved from its role as a vehicle of economic development (Diamond 1997) to the contemporary focus on sustainable tourism. Why is it, however, that tourism has been adopted so widely as a developmental option?

Why tourism?

As previously noted, few if any nations of the world have not become tourist destinations, and for many tourism has become an integral element of national development policy. In China, for example, tourism is a fundamental strand of the Western Region Development Strategy which aims to promote the socio-economic development of the country's western provinces, covering some 70 per cent of China's total land area (Zhang *et al.* 1999). In China there are 310 new hotels under construction, 210 of which are four- or five-star hotels. This rapid construction could result in more than 65,000 new rooms at a cost of $17 billion (Elegant 2006). Moreover, even for countries with a limited history of tourism, such as some of the former Soviet states in central Asia, it has become a development option of choice (see Box 1.2).

Box 1.2

Tourism and development in Kyrgyzstan

Located in central Asia, and bordered by China, Kazakhstan, Tajikistan and Uzbekistan, the Republic of Kyrgyzstan is a mountainous country of some 4.9 million people, 60 per cent of whom are ethnic Kyrgyz. In common with other former USSR states in the region, freedom from Soviet control since the early 1990s has, in some ways, proved to be a mixed blessing. On the one hand, it has enabled the country to adopt a policy of democratic reform; indeed, until the overthrow of President Akaev in 2005 (which, for the first time, brought it into the international limelight), Kyrgyzstan was seen as a stable, tolerant and multicultural country following a path towards economic development. On the other hand, political independence was, of course, accompanied by economic independence. No longer benefiting from Soviet investment and trade, the years following the end of Soviet domination have been characterized by socio-economic regression rather than development. Although being able to boast an adult literacy rate of 99.6 per cent, inherited from the Soviet era, Kyrgyzstan, with a per capita GDP of US$390, is one of the poorer countries in the world. About 55 per cent of the population live below the poverty line and over 80 per cent of the poorest people live in rural areas. The country also suffers from maternal and infant mortality rates that are more than twice the European average.

To address these challenges, the government has implemented a 'Comprehensive Development Framework for Kyrgyzstan', the purpose of which is two fold: to reduce poverty, and to enhance the country's international reputation. The latter is geared towards attracting international aid upon which the country's economy remains highly dependent. However, in order to achieve economic growth, tourism has been identified as an economic development priority and is, in fact, a named programme within the Comprehensive Development Framework. Underpinning the potential development of tourism is the country's natural resource base; more than 90 per cent of the country is over 1,000m altitude and, despite occupying just 0.13 per cent of the world's land area, it is claimed that Kyrgyzstan possesses around 2 per cent of the world's species of flora and more than 3 per cent of world fauna. The country also boasts Issyk-Kul Lake, the second largest high-altitude lake in the world. It was here that state-sponsored social tourism was promoted during the Soviet era and the lake remains fundamental to the country's tourism development plans. That is, one of the key priorities for tourism is the development of recreational tourism in the Issyk-Kul region targeted primarily at other Commonwealth of Independent States (CIS) countries. The international market is also a priority, with adventure tourism based on mountaineering, trekking and rafting being the main attractions on offer. Since the late 1990s, some progress has been made. For example, tourist arrivals have grown from around 35,000 in 1995 to about 211,000 in 2003, although the accuracy of the data is questionable and no distinction is made between regional (CIS) and international arrivals.

Unfortunately, however, just 4 per cent of the state budget is allocated to tourism and, as a result, the nascent private tourism sector in Kyrgyzstan receives little in the way of tourism development or promotional support. Little progress has been made in infrastructural development and, currently, there are just two international-standard hotels in the country. At the same time, there is little evidence to suggest that the country's official tourism development objectives have been translated into positive actions on the ground. Moreover, a recent NGO-sponsored community tourism project focusing on the development of nature and culture-based ecotourism among poorer rural communities has resulted in a variety of tensions emerging between different groups involved in tourism. Therefore, as is the case in many other developing countries, the realization of tourism's potential as an agent of economic and social development is dependent not only on appropriate policies but also on an economic and political climate that actively supports and coordinates the development of tourism.

Source: Palmer (2006).

In some instances, of course, tourism may represent the only realistic development path; that is, for some developing countries there is simply no other choice (Brown 1998: 59). More positively, however, the most compelling reason for adopting tourism as a development strategy is its potential contribution to the local or national economy, in particular to the balance of payments (Opperman and Chon 1997: 109). Many developing countries suffer severe balance of payment deficits and, as an export, tourism may represent a significant source of foreign exchange earnings. It is also widely considered to be a labour-intensive industry, and hence an effective source of employment in destination areas, whether direct employment in hotels, restaurants and so on, or indirect/informal employment (Farver 1984; Cukier and Wall 1994).

Beyond these basic economic drivers, however, a number of factors underpin the attraction of tourism as a development option.

- *Tourism is a growth industry*. As noted above, international tourism has demonstrated remarkable and consistent growth over the past half century, averaging 6.2 per cent annual growth since 1950. However, the rate of growth has been steadily declining. During the 1990s, for example, the average annual growth in tourist arrivals worldwide was 4.2 per cent, the lowest rate since the 1950s, although 2004 witnessed a remarkable growth of 10 per cent over the previous year (Table 1.5).

Table 1.5 International tourism arrivals and receipts growth rates, 1950–2000

Decade	Arrivals (average annual increase %)	Receipts (average annual increase %)
1950–60	10.6	12.6
1960–70	9.1	10.1
1970–80	5.6	19.4
1980–90	4.8	9.8
1990–2000	4.2	6.5

Source: Adapted from WTO (2005b).

Nevertheless, tourism remains one of the world's fastest growing industries and, globally, that growth is forecast to continue (See Plates 1.1 and 1.2).

Plate 1.1 ***Cuba, Varadero: Hotel construction.***

Plate 1.2 ***Tunisia, near Monastir: Hotel construction.***

Thus, tourism is seen essentially as a safe development option. However, it is important to point out that certain periods have witnessed low or even negative growth: in 2001, for example, global tourist arrivals were 0.5 per cent down from the year 2000 as a result of 9/11. More typically, though, the effects of external influences are more locally or regionally defined. The Indian Ocean tsunami in December 2004 had a devastating impact on the tourism industries of the Maldives, Sri Lanka and the Phuket area of Thailand, yet had little effect on total global arrivals (Sharpley 2005).

- *Tourism redistributes wealth.* Tourism is, in principle, an effective means of transferring wealth, either through direct tourist expenditure or international investment in tourism infrastructure and facilities from richer, developed countries. Through the promotion of domestic tourism, it also potentially redistributes wealth on a national scale – in India, for example, domestic tourism is significantly greater, in terms of tourist trips, than international tourism (Singh 2001). However, the gross value and net retention of tourist spending varies considerably from one destination to another – many destinations suffer 'leakages' whereby tourist expenditure finances the import of goods to meet tourists' needs – while overseas investment is conditioned by the global political economy of tourism (see Chapter 3).
- *Backward linkages.* Given the variety of goods and services demanded by tourists, from accommodation to local transport and souvenirs, tourism potentially offers more opportunities than other industries for backward linkages throughout the local economy, whether directly meeting tourists' needs, such as providing food to hotels (Telfer 1996) (see Plate 1.3), or through indirect links with, for example, the construction industry. Again, the extent to which such linkages can be developed depends upon a variety of factors, such as the availability of finance, the diversity and maturity of the local economy or the quality of locally produced goods.
- *Tourism utilizes natural, 'free' infrastructure.* The development of tourism is frequently based on existing natural or man-made attractions, such as beaches, wilderness areas or heritage sites (see Plate 1.4). Thus, tourism may be considered to have low 'start-up' costs when compared with other industries, since such resources are, in a simplistic sense, 'free'. Increasingly, however, attempts are being made to place an economic value on the use of these basic resources while, inevitably, costs are incurred in the protection, upkeep and management of all tourism resources.
- *No tourism trade barriers.* In many instances, individual countries or trading blocks, such as the European Union, impose restrictions of one form or another to protect their internal markets. In principle, international tourism faces no such trade barriers. That is, generating countries rarely place limitations on the right of their citizens to travel overseas, on where they visit and how much they spend (although travel advisories are one form of

Plate 1.3 ***Indonesia, Yogyakarta: A woman whose family owns and operates a small hotel in one of the tourist districts in the city is returning from a traditional market by three-wheel bicycle taxi after purchasing food for the hotel restaurant. She buys most of the food in traditional markets, thereby circulating tourist expenditures into the local economy.***

limitation on travel, as are possible travel visa costs and restrictions). Thus, destination countries have free and equal access to the international tourism market. However, the extent to which destinations can take advantage of this 'barrier-free' market is, of course, determined by international competition in general and by the structure and control of the international tourism system in particular. Indeed, as the following section demonstrates, developing countries continue to enjoy a relatively limited share of global tourist arrivals and receipts.

Plate 1.4 ***China: Tourists on the Great Wall of China, a UNESCO World Heritage Site.***
Source: Hui Di Wang.

Tourism demand

For current purposes, the demand for tourism may be considered from two perspectives:

1. Historical and contemporary patterns and flows of international tourism (i.e. statistical data).
2. Transformations in the nature of the demand for tourism (i.e. changes in styles of tourism).

International tourism trends and flows

Reference has already been made to the remarkable and continuing growth of international tourism since the 1950s. This growth has been restricted at times by a variety of events, including international conflicts such as the 1991 Gulf War, the oil crises of the 1970s, global economic recession in the early 1980s and 1990s, and a variety of specific events such as health scares, natural disasters and, of course, terrorist activity. Only rarely, however, have global, as opposed to regional, arrivals experienced a downturn (Table 1.6).

Importantly, the global growth of international tourism has not been equitable. That is, not all parts of the world have experienced similar growth rates, and international tourism is still largely dominated by the industrialized world, with major tourist flows being primarily between the more developed nations and, to a

Table 1.6 International tourist arrivals and receipts, 1950–2004

Year	Arrivals (million)	Receipts (US$bn)	Year	Arrivals (million)	Receipts (US$bn)
1950	25.3	2.1	1994	535.8	356.0
1960	69.3	6.9	1995	550.4	404.6
1965	112.9	11.6	1996	580.2	438.8
1970	165.8	17.9	1997	601.5	442.9
1975	222.3	40.7	1998	621.4	445.2
1980	286.5	105.4	1999	643.3	455.0
1985	328.8	118.0	2000	687.3	473.4
1990	455.9	264.1	2001	684.1	459.5
1991	461.1	277.9	2002	702.6	472.2
1992	502.2	317.1	2003	690.7	523.1
1993	515.3	322.9	2004	760.0	n/a

Source: Adapted from WTO (2004a, 2005b).

lesser extent, from developed to less developed countries. As Table 1.7 shows, about half of all international arrivals can be accounted for by just ten nations, collectively attracting 49 per cent of total global arrivals in 2003.

Not surprisingly, a similar pattern is evident in terms of tourism receipts (Table 1.8). The USA has long been the greatest beneficiary of international tourism while, in 2003, almost 53 per cent of global tourism receipts were earned by the top ten countries. Of the top ten most popular destinations for international tourism, six are also ranked within the ten largest tourism generators (Table 1.9).

The pattern of international tourist flow is reflected in regional shares of tourist arrivals and receipts. Europe has long received the greatest proportion of international arrivals, although, as is evident from Table 1.10, despite continuing to enjoy an annual increase in international arrivals, its share of the global market has been shrinking steadily.

Table 1.7 World's top ten international tourism destinations, 2003

		Arrivals (millon)	Share of total (%)
1	France	75.0	10.9
2	Spain	51.8	7.5
3	United States	41.2	6.0
4	Italy	39.6	5.7
5	China	33.0	4.8
6	UK	24.7	3.6
7	Austria	19.1	2.8
8	Mexico	18.7	2.7
9	Germany	18.4	2.7
10	Canada	17.5	2.5

Source: Adapted from WTO (2004a).

Table 1.8 World's top ten international tourism earners, 2003

		Receipts ($billion)	Share of total (%)
1	United States	64.5	12.3
2	Spain	41.8	8.0
3	France	37.0	7.1
4	Italy	31.2	6.0
5	Germany	23.0	4.4
6	UK	22.8	4.3
7	China	17.4	3.3
8	Austria	14.1	2.7
9	Turkey	13.2	2.5
10	Greece	10.7	2.0

Source: Adapted from WTO (2004a).

Table 1.9 World's top ten international tourism generators by expenditure

		Expenditure ($ billion)	Share of total (%)
1	Germany	64.7	12.4
2	United States	56.6	10.8
3	UK	48.5	9.3
4	Japan	29.0	5.5
5	France	23.6	4.5
6	Italy	20.5	3.9
7	China	15.2	2.9
8	Netherlands	14.6	2.8
9	Canada	13.3	2.5
10	Russian Fed.	12.9	2.5

Source: Adapted from WTO (2004a).

Table 1.10 Percentage share of international tourist arrivals by region, 1960–2003

	Africa	Americas	EAP	S. Asia	Europe	M. East
1960	1.1	24.1	1.1	0.3	72.6	0.9
1970	1.5	25.5	3.2	0.6	68.2	1.1
1980	2.6	21.6	7.4	0.8	65.6	2.1
1990	3.3	20.4	12.0	0.7	61.6	2.2
1995	3.6	19.8	14.8	0.8	58.6	2.5
2000	4.0	18.6	15.9	0.9	57.1	3.5
2001	4.1	17.6	16.6	0.9	57.1	3.5
2002	4.2	16.3	17.9	0.8	56.9	3.9
2003	4.5	16.4	16.3	0.9	57.7	4.2

Source: Adapted from WTO (2004a).

Conversely, the East Asia Pacific (EAP) region has enjoyed spectacular growth in tourist arrivals, overtaking the Americas in 2002 to become the world's second most visited region that year. Indeed, during the 1990s, annual arrivals in the region doubled while receipts grew by 121 per cent, both figures being twice the global rate. The main destinations in EAP are China, Hong Kong, Thailand, Malaysia and Singapore, although interestingly, newer destinations in the region, including Vietnam, Cambodia, Myanmar, Laos and Polynesia, have successfully developed their tourism sectors. Much of this growth can be accounted for by intra-regional travel. Other regions of the world have also increased their share of the global tourism market. Annual international arrivals in the Middle East more than doubled during the 1990s, with Egypt, Bahrain, Jordan and, in particular, Saudi Arabia and the UAE enjoying rapid growth.

Thus, overall there has been a gradual shift away from the traditional destinations of Europe and North America to other regions of the world. Indeed, according to the WTO (2005c), during the period 1995–2002, over 30 nations enjoyed tourism growth at more than double the global rate. All of these were developing countries although, collectively, the developing world attracts only about one-third of total international tourism receipts. Nevertheless, the arrivals/receipts data mask the importance of tourism to the economies of many developing countries; that is, although their arrivals figures tend to be relatively low, the contribution of tourism to the national economy is high. For example, data provided by the World Travel and Tourism Council (WTTC 2004) demonstrate that, in 25 countries, the tourism economy contributes more than 25 per cent of GDP – the most tourism-dependent nation is the British Virgin Islands, where tourism contributes 95 per cent of GDP. All 25 are developing countries (and small island states) with tourism industries of different magnitudes. Therefore, a destination's share of global arrivals is less significant, in developmental terms, than the relative importance of tourism to the local economy.

The nature of tourism demand

The dramatic growth and spread of international tourism over the past 50 years has been driven by a variety of factors. Typically, increases in wealth and free time, and technological advances in transport, are considered to be the principal influences on the development of tourism, although political change has, in more recent years, been a significant factor in the emergence of new destinations in the developing world, particularly in the former Soviet republics. At the same time, the emergence of a sophisticated travel industry, initially providing the 'package holiday' to mass markets inexperienced in international travel, and, more recently, developing products ranging from low-cost/no-frills flights to Internet-based 'dynamic packaging' (whereby tourists can construct their own package holidays by booking flights, accommodation, car hire and other services from different suppliers) has fuelled the growth of tourism.

However, of particular relevance to developing countries, the nature of tourism demand has also changed and evolved over the past 20 years. Although the standardized, sun-sea-sand package holiday remains the most popular form of tourism, at least among tourists from the developed Western world, there has been a dramatic growth in demand for the more individualistic, active/participatory forms of tourism providing a broader or more fulfilling experience. An example of this is the growth in demand for cultural tourism, adventure tourism, heritage tourism, ecotourism and, more generally, an expansion of long-haul tourism. This, in turn, suggests that tourists have become more experienced, discerning and quality-conscious, and more adventurous in the practice of consuming tourism experiences.

These issues are considered in more detail in Chapter 6 but the important point is that the demand for tourism has allegedly been characterized by the emergence of the so-called 'new tourist' (Poon 1993). Tourists are now considered to be more flexible, more environmentally sensitive, more adventurous and inclined to seek out meaningful experiences and, hence, travelling increasingly to different, more distant, untouched and exotic or new destinations. Whether or not this can be explained by the existence of the 'new' tourist remains debatable but there is no doubt that changing tastes in tourism demand represent a vital opportunity for developing countries; the challenge lies in the extent to which they are able to harness this potential means of development.

Tourism supply

The ability of tourists to enjoy travel or vacation experiences is largely dependent upon the multitude of organizations that collectively supply the goods and services required by tourists. Typically, these include transport both to and within a destination, accommodation, food and drink, entertainment (attractions or activities) and shopping, together with associated services such as insurance and finance. At the same time, public sector organizations frequently support the supply of tourism through, for example, regional or national marketing or the provision of information services for visitors. In some cases, such as cruise holidays, the core elements of the tourism product (transport, accommodation, sustenance, entertainment) are supplied collectively, while the success of tour operators is based on the packaging of different components into a single product – the package holiday. In other cases, of course, tourists organize their holidays independently but, nevertheless, the key elements remain the same.

However, the manner in which tourism is supplied may have a significant impact on the contribution of tourism to development. In other words, although the tourism industry supplies similar components in the provision of tourist experiences, the characteristics of that supply may vary considerably in terms of scale, nature and control/ownership. At one extreme, for example, a destination may be characterized by standardized, large-scale, mass tourism development largely owned or controlled by overseas interests, potentially limiting tourism's developmental contribution and reducing the destination to a state of dependency (Britton 1982). At the other extreme, tourism development may be small scale, appropriate to the local environment and locally owned and controlled, thereby potentially optimizing the developmental benefits to the local community. Indeed, such extremes may represent the two ends of a tourism development continuum.

At one end lies, arguably, the least desirable situation whereby a destination becomes, in a sense, an annex of the wealthy developed countries, with tourism representing a modern form of colonial exploitation; at the other (ideal) end, tourism contributes effectively to local sustainable development.

In reality, most tourism developments fall somewhere between these two poles, while the relationship between the characteristics of tourism supply and its contribution to development remains more complex. For example, over the past 25 years, the Dominican Republic has built an economically successful tourism sector, attracting some 2.8 million tourists annually and now contributing, directly and indirectly, 25 per cent of the country's GDP and some 300,000 jobs, or 22 per cent of employment. However, that success has resulted primarily from the development of what many would consider to be 'unsustainable' all-inclusive mass tourism resorts. Moreover, in many instances the contribution of tourism to development is enhanced not by the scale and ownership of tourism resources, but by innovative schemes implemented by the international tourism industry or other relevant organizations. One such example is the Travel Foundation, a UK-based charity that works with the outbound travel industry to manage tourism more sustainably, as in the case of the SAVE (Save Abandoned Villages and their Environment) project in Cyprus. Here, leading tour operators worked with local partners to develop excursions (for package tourists) to rural villages on the island, representing a valuable source of income to local businesses and at the same time providing visitors with an experience of traditional Cypriot culture they might not otherwise enjoy (see www.thetravelfoundation.org.uk).

The important point is that simplistic models of appropriate/inappropriate tourism supply cannot meet the diversity of developmental contexts. In other words, local social, economic or political structures and developmental needs may dictate the nature of tourism supply. Thus, while the Himalayan kingdom of Bhutan has long pursued a restrictive policy on tourism development, strictly controlling visitor numbers to maintain the cultural integrity of the country, other destinations, such as Cancún in Mexico, have followed a more expansionist tourism development policy based on foreign investment in mass tourism enterprises (see Clancy 1999; Telfer 2002b). Nevertheless, an integrated, community-focused approach to tourism supply is widely considered to be the most appropriate form of tourism development, reflecting the contemporary dominance of the sustainable tourism development paradigm (Mann 2000).

Tourism and development

Just as development theory in general has evolved over time, so too have approaches to tourism development in particular. Moreover, the evolution of tourism theory has also largely reflected the evolution of development theory as described earlier in this chapter, although any relationship between the two is not always clear. Indeed, until recently (see e.g. Sharpley and Telfer 2002; Mowforth and Munt 2003) there has been little interaction between the two fields of study, while successive approaches to tourism development have been linked primarily to concerns over its socio-environmental impact (Dowling 1992) rather than an understanding of tourism and development processes. Nevertheless, based on an analysis of tourism literature, Jafari (1989) identifies four stages or 'platforms' of tourism theory which, to an extent, parallel development theory:

- *Advocacy*. During the 1960s, tourism was viewed as a positive vehicle of national and international development. Reflecting modernization theory, its potential was considered to lie in its contribution to economic growth, this being measured by indicators such as income and employment generation and the multiplier effect. In short, tourism was seen as an effective developmental growth-pole, as indeed it continues to be seen in some contexts.
- *Cautionary*. From the late 1960s onward, concern was increasingly expressed over the negative environmental and socio-cultural consequences of tourism that were resulting from both the scope, scale and rapidity of tourism development, and the emerging political economy of tourism which reflected the dependency paradigm. Thus, at this stage tourism theory was concerned with understanding tourism's impacts on destination environments and societies, with a particular focus on centre–periphery dependency models (Høivik and Heiberg 1980).
- *Adaptancy*. As a response to the preceding antithetical positions on tourism development, the 1980s witnessed the emergence of alternative, though idealistic, approaches to tourism. Variously referred to as 'green', 'appropriate', 'responsible', 'soft' or 'alternative' tourism, these approaches attempted to transpose the principles of alternative development on to tourism, proposing appropriately scaled, locally owned and controlled development, with the community as the primary instigators and beneficiaries of tourism. To an extent, this approach remains reflected in the contemporary (although contentious) concept of ecotourism.
- *Knowledge*. As greater knowledge of tourism's developmental processes has emerged, the idealistic ambition of alternative tourism has been overtaken by a broader approach to tourism development that attempts to embrace the principles and objectives of sustainable development. Thus tourism, as a specific developmental vehicle, has aligned itself with the contemporary development paradigm although, as considered in detail in Chapter 2, sustainable tourism development has proved to be problematic both in its practical implementation and in its acceptance by the many developing countries which view the concept as evidence of continuing Western imperialism.

Approaches to tourism development, then, have evolved over time from traditional, modernist economic growth models through to sustainable approaches that attempt to balance tourism as a profit-driven, resource-hungry activity with the developmental needs of destination environments and communities. The latter have been collectively referred to by the World Travel and Tourism Council as 'New Tourism' (WTTC 2003), an approach that demands an effective, long-term partnership between the public and private sectors, though, interestingly, no explicit reference is made to community participation. In other words, the most recent approaches to tourism development, such as 'pro-poor tourism' (www.propoortourism.org.uk), have shifted the emphasis towards responsible activities on the part of the international tourism industry.

This, however, will do little to alleviate the dilemma facing tourism destinations in the developing world. That is, to many developing countries, tourism represents a potentially valuable development option, yet it is associated with a variety of costs or impacts, from environmental degradation to dependency on international corporations. At the same time, adopting a policy of larger scale tourism development may provide greater economic benefits in terms of income and employment, though with potentially greater impact; conversely, the adoption of smaller scale, appropriate tourism may lessen the impact, but may also result in a reduced developmental contribution.

The challenge for developing countries is therefore to seek ways of resolving this tourism development dilemma, and it is with this issue that this book is concerned. The next chapter reviews sustainable tourism development, focusing in particular on contemporary debates that challenge or question the validity of the concept, and highlighting issues that are then addressed in subsequent chapters. These include the relevance and influence of globalization on tourism in developing countries (Chapter 3), and an examination of the tourism planning and development process and the potential contributions to development by selected forms of tourism (Chapter 4). Community involvement in tourism remains a central tenet of sustainable tourism development, and Chapter 5 explores the community response to tourism along with various forms of community-oriented tourism. Chapter 6 considers tourism development from the perspective of the tourist, addressing in particular the implications of transformations in the consumption of tourism for sustainable tourism development. Chapter 7 highlights the impact of tourism development and explores contemporary techniques for minimizing such impact. Finally, Chapter 8 draws together the themes and debates raised throughout the book, and presents a tourism–development dilemma framework that illustrates the complexity of often interconnected forces at work in using tourism as a development tool. The chapter argues that there is a development imperative, and a sustainable development imperative. However, it is important to realize the challenges of implementing the ideals of sustainable development in the context of the realities of the tourism industry in developing countries.

Discussion questions

1. **Why is tourism selected as a development tool by so many developing nations?**
2. **What are the structural dimensions in developing countries that relate to underdevelopment, and what role can tourism play in addressing these problems?**
3. **Is there a new emerging tourism market?**
4. **How have development approaches to tourism changed over time?**

Further reading

Mowforth, M. and Munt, I. (2003) *Tourism and Sustainability: Development and New Tourism in the Third World*. (2nd edn) London: Routledge. [This book is essential reading. It provides a comprehensive and detailed critique of contemporary approaches to tourism development within the framework of development, globalization, power relations and sustainability. In particular, it challenges the widespread optimism for new approaches to tourism, questioning the role of tourism within the global political economy.]

Reid, D. (2003) *Tourism, Globalization and Development: Responsible Tourism Planning*. London: Pluto Press. [Questioning the potential contribution of corporate-dominated tourism development to the social and economic development in less developed countries, this book provides a clear analysis of the relationship between tourism and development and, in particular, highlights the benefits of community-led tourism planning and development.]

Sharpley, R. and Telfer, D. (2002) *Tourism and Development: Concepts and Issues*. Clevedon: Channel View Publications. [The first section of this book provides a detailed introduction to the relationship between tourism, development and development theories. This is followed by both a comprehensive analysis of key issues relevant to tourism's contribution to development, and also a critique of contemporary challenges and barriers to the achievement of sustainable development through tourism.]

Websites

Up-to-date research reports and studies about the implementation and benefits of pro-poor tourism: www.propoortourism.org.uk.

A charitable organization that encourages the outbound travel industry to manage tourism more sustainably: www.thetravelfoundation.org.uk.

Information on development processes and progress in the developing world in general, and on the activities and programmes of the United Nations Development Programme in particular (including the UNDP's annual Human Development Report): www.undp.org.

A useful source of information regarding economic and social development goals, processes and statistical data in less developed countries: www.worldbank.org.

The World Tourism Organization (UNWTO), an essential information source for contemporary international tourism trends statistics, global tourism policies and tourism development guidelines: www.world-tourism.org.

The World Travel and Tourism Council (WTTC), a valuable source of information and statistical data, in particular via its Document Resource Centre which provides access to all published WTTC documents, research and policy statements: www.wttc.org.

2 Tourism and sustainable development

Learning objectives

When you have finished reading this chapter, you should be able to:

- **Understand the evolution, principles and objectives of the concept of sustainable development;**
- **Evaluate the key debates surrounding the definition, implementation and measurement of sustainable tourism development;**
- **Identify and assess contemporary approaches to sustainable tourism development;**
- **Understand the links between sustainable development, globalization and political economy.**

In 1987, the World Commission on Environment and Development (WCED) published its report, *Our Common Future* (WCED 1987). Better known, perhaps, as the Brundtland Report, this document argued that the most effective way to 'square the circle of competing demands for environmental protection and economic development' (Dresner 2002) was through the adoption of a new approach, namely sustainable development. Although the WCED did not, in fact, coin the term – the World Conservation Union had referred to sustainable development in its earlier *World Conservation Strategy* (IUCN 1980) – the report brought the concept to the attention of a much wider audience and, since the late 1980s, sustainable development has dominated global development policy. Not only has it provided the focus for a number of major international events, from the 1992 'Earth Summit' in Rio de Janeiro to the World Summit on Sustainable Development ('Rio +10') in Johannesburg in 2002, but also innumerable organizations in the public, private and voluntary sectors and at the international, national and local levels have embraced its principles and objectives. It is also the focus of Agenda 21, the action plan for sustainable development that came out of the 1992 'Earth Summit'.

Reflecting the emergence and growing acceptance of sustainable development in general, the concept of sustainable *tourism* development in particular also came to prominence towards the end of the 1980s, subsequently achieving 'virtual global endorsement as the new [tourism] industry paradigm' (Godfrey 1996: 60). In both the public and private sectors, a plethora of policy documents, planning guidelines, statements of 'good practice', corporate social responsibility statements, case studies, codes of conduct for tourists, and other publications have since been produced, all broadly concerned with the issue of sustainable tourism development. Moreover, the importance of the concept was highlighted at the 2002 World Summit on Sustainable Development (UNEP/WTO 2005) while, in that same year, the city of Quebec hosted the World Ecotourism Summit – ecotourism arguably being a specific manifestation of sustainable tourism. In short, for the past 15 years tourism policy and planning has, by and large, been driven by the principles and objectives of sustainable tourism development, although, as we shall see, the extent to which those objectives have been achieved remains debatable.

To a great extent, the widespread support for the concept of sustainable tourism development is not surprising. As noted in Chapter 1, the initial euphoria over the developmental potential of international tourism had, by the late 1960s, been replaced by a more cautionary approach as increasing numbers of commentators drew attention to the potentially destructive effects of tourism development. Initially, concerns were voiced by the 'Limits to Growth' school who reflected the contemporary criticism of unbridled economic growth (Schumacher 1974; Andersen 1991) and called for restraint in the development of tourism (Mishan 1969; Young 1973). More specific studies of tourism's consequences followed in the late 1970s and early 1980s (Turner and Ash 1975; Smith 1977; de Kadt 1979; Mathieson and Wall 1982) and, by the 1990s, tourism (or, more specifically, mass tourism) was being described in almost apocalyptic terms as a 'spectre . . . haunting our planet' (Croall 1995: 1). As a result, the principles of sustainable tourism development, which address many (often justifiable) concerns and criticisms of mass tourism, were widely adopted at national and destinational levels, as well as by certain sectors of the travel and tourism industry.

However, sustainable tourism development remains a highly controversial concept, as indeed does its parental paradigm of sustainable development. Specifically, it is seen as divisive, polarizing the debate between sustainable or 'good' forms of tourism and unsustainable, mass (or 'bad') forms of tourism, as well as being an inflexible blueprint that cannot be adapted to different tourism developmental contexts. More generally, not only have many questioned the extent to which tourism can be mapped on to the broad principles and objectives of sustainable development (Sharpley 2000; Berno and Bricker 2001), but it is also accepted that there are few, if any, examples of 'true' sustainable tourism development in practice. Moreover, there is recent evidence to suggest that idealistic support for the concept is on the wane, and the adoption of a more pragmatic and 'responsible' approach to tourism development is being manifested in specific policies, such as pro-poor tourism.

Nevertheless, it is still recognized that 'tourism is in a special position in the contribution it can make to sustainable development' (UNEP/WTO 2005: 9). In other words, for many countries, tourism represents the principal, sometimes the only, route to development, and thus there is a need to ensure it is developed in such a way that its contribution to a destination's sustainable development is optimized. The purpose of this chapter, therefore, is to explore the principles and objectives of sustainable development before going on to consider the debates surrounding the concept of sustainable tourism development and the contemporary challenges it faces. Thus the first question to be addressed is: What is sustainable development?

Sustainable development: towards a definition

Despite the widespread acceptance of sustainable development, there remains a lack of consensus over the actual meaning of the term. That is, although it is relatively easy to define what sustainable development is *not*, saying what it *is* has proved to be more problematic. It means different things to different people and is applied to innumerable contexts (including, of course, tourism). As a result, definitions abound – it has been observed, for example, that by the early 1990s over 70 definitions of sustainable development had been proposed (Steer and Wade-Gery 1993)! Nevertheless, the most popular and enduring remains the Brundtland Report's definition as 'development that meets the needs of the present without compromising the ability of future generations to meet their own needs' (WCED 1987: 48), although this is widely regarded as, at best, vague and, at worst, meaningless.

It is generally agreed, however, that sustainable development represents a 'meeting point for environmentalists and developers' (Dresner 2002: 64). In other words, sustainable development may be thought of as a combination of two processes, namely development and sustainability (although the latter term, somewhat confusingly, is often used interchangeably with sustainable development). Such an approach does, perhaps, oversimplify matters. That is, both 'development' and 'sustainability' are open to interpretation – for example, it has long been recognized that environmental sustainability can be considered from either an ecocentric (strong sustainability) or technocentric (weak sustainability) perspective. Nevertheless, for the purposes of this book, dividing sustainable development into its two constituent elements provides a useful basis for exploring the evolution of the concept.

'Development' in general, and transformations in the meaning or interpretation of development in particular, have already been addressed in Chapter 1. It is, then, necessary to consider the other half of the sustainable development 'equation', namely sustainability.

From conservation to sustainability

Just as development thinking has evolved from the narrow, classical economic growth perspective to the broader, alternative development and recent diversity of approaches, so too has the nature of environmental concern, or environmentalism, evolved through a number of stages, from conservation to sustainability.

Concern about human impact on the environment can be traced back almost to the beginning of civilization while societies have suffered throughout history (and continue to suffer) from a variety of environmental problems, such as over-population, resource depletion and pollution (McCormick 1995). However, it was not until around the 1850s that a formal and organized conservation movement began to emerge, signalling the advent of contemporary environmentalism. Influenced by a number of factors, including urban and industrial development, greater interest in and knowledge of the natural world, and an evolving amenity movement demanding greater access to natural spaces, a large number of organizations were established that sought either to protect particular species, such as Britain's Royal Society for the Protection of Birds, founded in 1889, or to preserve natural areas, such as the USA's Sierra Club, founded by John Muir in 1892. All these organizations had a common aim, and one very much in opposition to the prevailing modernist–technocentric belief in the domination and exploitation of nature; that is, to promote the conservation (and public enjoyment) of natural resources.

This focus on conservation remained predominant until the mid-twentieth century. Moreover, despite the creation of international organizations such as the International Union for the Conservation of Nature (IUCN), now known as the World Conservation Union, the scope of environmental concern was usually local and rarely transcended national boundaries or interests. From the 1960s, however, environmentalism became a popular ideology with a set of preoccupations that went far beyond the specific concerns of protecting threatened natural areas and species. Rather than focusing simply on resource depletion, the actual scientific, technological and economic processes upon which human progress was previously seen to depend were also questioned. For example, Rachel Carson's *Silent Spring* (1962) was specifically concerned with the misuse of synthetic pesticides, yet it is seen by many as a landmark event in the history of modern environmentalism. Similarly, Hardin's *Tragedy of the Commons* (1968) described in simple terms how individual over-consumption of a limited natural resource eventually brings ruin to all.

At the same time, it was acknowledged that the by-products of industrialization, the so-called 'effluence of affluence', did not respect national boundaries: environmental problems, such as air and water pollution, frequently originated in one country but adversely affected another. Influenced by Boulding's notion of 'spaceship earth' in the mid-1960s, environmentalism took on an international dimension. The earth became viewed as a closed system with finite resources and a limited capacity to absorb waste and, as a result, the threat to the world's environment came to be seen as a global crisis. Thus, it was no coincidence that the motto of the United Nations Conference on the Human Environment (UNCHE) in 1972 was 'Only One Earth'.

Environmentalism, then, differed from earlier conservation in two respects. First, it addressed the entire human environment, embracing not only resource problems, such as acid rain, deforestation and whaling, but also the underlying technological and economic processes that led to such problems. International travel, particularly air travel, has come to be seen by many as one such process that makes a significant contribution to environmental degradation and what has become the dominant political–environmental issue of today, namely climate change. Second, it became more overtly political and activist – environmentalists turned their attention to social and political issues and embraced other social movements, including the anti-war, anti-consumerism and civil rights movements that flourished in the 1960s and 1970s. Nevertheless, their main concern was (and remains) the earth's capacity to support human existence, providing the foundation for what is now referred to as 'sustainability'.

In essence, sustainability is based on the notion that the human economic system of production and consumption is a subsystem of the global ecosystem; this, in turn, is the source of all input into the economic subsystem and the sink for all its wastes (Goodland 1992). The global ecosystem's source and sink functions have a finite capacity to supply respectively the needs of production/consumption and absorb the wastes resulting from these processes. Thus, the variables in the equation become:

1. The rate at which the stock of natural (non-renewable) resources is depleted relative to the development of substitute, renewable resources;
2. The rate at which waste is deposited back into the ecosystem relative to the assimilative capacity of the environment;
3. Global population levels and per capita levels of consumption.

Sustainability is concerned with maintaining a balance between these variables, the potential for which is, to a greater extent, dependent on the effective management (or, perhaps, transformation) of the political, technological, economic and social institutions that determine the production and consumption processes. Each variable is of equal importance to the achievement of sustainability, although, following the Kyoto Treaty of 1997, most attention is currently focused on global warming and climate change. Indeed, at a UN conference in Montreal in late 2005, all but one member state (the USA) agreed to speed up the measures agreed at Kyoto, reflecting widespread concern for an issue that is, as we shall see shortly, of particular relevance to tourism.

Sustainable development: principles and objectives

During the 1970s, attention was focused primarily on limiting growth in order to reduce what were considered to be excessive demands upon the global ecosystem. The IUCN's *World Conservation Strategy* in 1980, referred to above, introduced the term 'sustainable development' and made some attempt to integrate development with conservation, although its emphasis was firmly on the latter. Thus it was left to the Brundtland Report to combine development and environmental issues within a global strategy for sustainable development. This was subsequently followed by the lesser known, but no less important, *Caring for the Earth* document (IUCN 1991). Whereas the Brundtland Report's strategy for sustainable development was based very much on a return to neo-classical economic growth (Reid 1995) – that is, it espoused traditional economic growth-based development as the foundation for sustainable development – *Caring for the Earth* gave precedence to a strategy for 'sustainable living', the emphasis being on the adoption of sustainable lifestyles and, essentially, a transformation in people's attitudes towards consumption practices. Together, these reports provide a framework for identifying the key principles and objectives of sustainable development and, indeed, the prerequisites for its achievement.

Building on the basis that any form of development should occur within environmental limits, sustainable development should be guided by the following principles:

- *Holistic perspective*: Development and sustainability are global challenges;
- *Futurity*: The emphasis should be on the long-term future;
- *Equity*: Development should be fair and equitable both within and between generations.

As already observed, while the Brundtland Report recognized the desirability of meeting basic needs, it argued for economic growth as a prerequisite for sustainable development. Conversely, *Caring for the Earth* argued for a fundamental transformation in people's approach to the environment in general and consumption in particular. However, both are important elements of sustainable development. A certain level of wealth is necessary to underpin development, yet, globally, environmental pressures are not resource problems – they are human problems (Ludwig *et al.* 1993). Therefore, as summarized in Table 2.1, in order to meet the developmental and environmental objectives of sustainable development, a number of requirements must be satisfied.

The overall objectives of sustainable development, then, may be seen as:

- *Environmental sustainability*: The conservation and effective management of resources;
- *Economic sustainability*: Longer term prosperity as a foundation for continuing development;
- *Social sustainability*: With a focus on alleviating poverty, the promotion of human rights, equal opportunity, political freedom and self-determination.

It is important to note, of course, that this discussion of the principles and objectives of sustainable development reflects a dominant operational/managerialist perspective. That is, it is concerned with principles and processes deemed necessary to preserve the world's resources. However, development is also an inherently political process – differing perspectives on development may reflect differing political ideologies. For example, neoliberalism, which espouses both personal freedom and development based on free-market economic growth, runs counter to the radical ecological movement that sees the resolution of environmental problems lying with intervention in and restrictions on individual freedoms (Martell 1994). In a similar sense, socialist political economy proposes that sustainability may be achieved by reducing the influence of capitalism and the drive for profit, which itself requires a transformation in relations of production with

Table 2.1 Sustainable development: principles and objectives

Fundamental principles:	• *Holistic approach*: Development and environmental issues integrated within a global social, economic and ecological context. • *Futurity*: Focus on long-term capacity for continuance of the global ecosystem, including the human subsystem. • *Equity*: Development that is fair and equitable and which provides opportunities for access to and use of resources for all members of all societies, both in the present and future.
Development objectives:	• Improvement of the quality of life for all people: education, life expectancy, opportunities to fulfil potential. • Satisfaction of basic needs: concentration on the nature of what is provided rather than income. • Self-reliance: political freedom and local decision-making for local needs. • Endogenous development.
Sustainability objectives:	• Sustainable population levels. • Minimal depletion of non-renewable natural resources. • Sustainable use of renewable resources. • Pollution emissions within the assimilative capacity of the environment.
Requirements for sustainable development:	• Adoption of a new social paradigm relevant to sustainable living. • Biodiversity conservation. • International and national political and economic systems dedicated to equitable development and resource use. • Technological systems that can search continuously for new solutions to environmental problems. • Global alliance facilitating integrated development policies at local, national and international levels.

Sources: Streeten (1977); Pearce *et al.* (1989); WCED (1987); IUCN (1991).

collective ownership. Other political perspectives on development and sustainability include both the Marxist approach (which some claim contributes positively to environmentalism by emphasizing people's need to interact with their natural environment to overcome a sense of alienation and also achieve personal freedom and fulfilment), and an eco-feminist perspective. The latter proposes that values and traits associated with femininity (such as an emphasis on interrelationships, balance and natural processes) may provide the foundation for sustainability (Martell 1994).

Sustainable development: from principle to practice

As Dresner (2002: 67) observes, 'sustainable development is not such a vague idea as it is sometimes accused of being'. However, it remains a controversial concept, largely as a result of problems in its operationalization and measurement. Moreover, as noted in the previous section, it is also susceptible to varying ideological interpretations. In other words, both putting the principles of sustainable development into practice and assessing their effectiveness has proved to be a difficult task. It is beyond the scope of this chapter to review fully the debate (see e.g. Reid 1995), although it is important to highlight the key questions or criticisms, as these may equally be levelled at the concept of sustainable tourism development in particular.

The principal difficulty facing the operationalization of sustainable development is the long-recognized ambiguity and inherent contradictory nature of the concept (Redclift 1987). That is, the twin objectives of 'development' and 'sustainability' are considered by many to be an oxymoron. How can development (necessitating resource exploitation) be achieved at the same time as sustainability (minimizing resource depletion)? Moreover, different interpretations of the phrase itself, and its two constituent elements, complicate matters further. Is sustainable development literally development that can be sustained, giving precedence to development (however defined), or is it development that is restricted by environmental sustainability (again, however defined)? Indeed, even if objective assessments of sustainability could be made, development inevitably remains a subjective concept. Thus, for example, while many consider tourism in Bhutan to be environmentally sustainable, the extent to which it is contributing to the country's sustainable development remains debatable (see Box 2.1).

Box 2.1

Tourism and sustainable development in Bhutan

The Himalayan Kingdom of Bhutan remained relatively isolated from the rest of the world until the 1960s. Indeed, it was not until the mid-1970s that small numbers of tourists began to visit the country although, at that time, both gaining the necessary visas and actually travelling to Bhutan (the only road access being via India) was a complex and lengthy process. Consequently, relatively few tourists chose to visit Bhutan and it was not until the opening of the international airport in 1983, and the subsequent extension of the runway in 1990, that greater numbers (in Bhutanese terms) began to arrive.

From the outset, tourism in Bhutan was introduced with the principal objective of generating foreign exchange earnings. Given its mountainous terrain, the country has few natural resources upon which to build its economy (although the generation and export of hydroelectric power to neighbouring India accounts for some 12 per cent of GDP). Agriculture and forestry are major activities, yet only 7 per cent of the land area is suitable for cultivation. Moreover, although the country is often portrayed as a Shangri-La, remote mountain communities follow a harsh existence: there is a lack of access to services, 80 per cent of the population are subsistence farmers and many rural households have an income of less than $1 per day. Indeed, 32 per cent of the population live in poverty. However, the country possesses a rich and diverse cultural and natural heritage, providing the principal attraction for tourists. It is home to a number of rare Himalayan species, boasts spectacular scenery, and has a vibrant culture manifested in distinctive architecture and spectacular festivals.

Despite its potential for high earnings from tourism, development in Bhutan has, to date, focused on high-yield but low-impact tourism. In order to protect its cultural and natural heritage for local communities and visitors alike, the scale and nature of tourism has been tightly controlled through indirectly restricting the number of arrivals: charging a minimum day rate (since 1991, this has been US$200 in high season, and US$165 in low season) and regulating all aspects of the local tourism industry, from the number of local tour operators to the supply of accommodation. Initially just 200 international tourists a year were permitted entry, but this figure has gradually risen over the years. In 1992, 2,850 arrivals were recorded; by 2000, this had risen to 7,559, generating gross receipts of US$10.5 million (7 per cent of GDP). Over 9,000 arrivals were recorded in 2004 and the objective is to increase this to 15,000 by 2007.

The activities of tourists within the country are also strictly controlled – although the industry has been 'privatized', every aspect of tourism is regulated by the Department of Tourism – the purpose being to minimize the negative impact of tourism. However, while this has resulted in Bhutanese tourism being widely seen as an example of successful sustainable tourism practice, questions must be raised over the extent to which tourism is genuinely contributing to sustainable development. In particular:

- Regulation restricts the extent to which local people can access the industry and, hence, benefit financially;
- Tourism to Bhutan is highly seasonal, impacting on permanent employment and income levels;
- Tourism activity is limited to a few western and central valleys; much of the country does not receive tourists and therefore does not share directly the benefits of tourism;
- Local community involvement in tourism is extremely limited;
- Despite relatively low numbers of tourists, the natural environment has suffered significant impacts, particularly deforestation, the erosion of delicate vegetation and the creation of 'garbage trails';
- An increase in the number of licensed tour operators to facilitate greater numbers of tourists has led to price competition among operators, thereby challenging the high yield–low volume policy.

Thus, as tourism plays an increasingly important role within the Bhutanese economy (tourism's share of GDP is forecast to rise to 25 per cent by 2012), appropriate policies will be needed to ensure that its contribution to sustainable development is improved.

Sources: Dorji (2001); Brunet *et al.* (2001).

Further criticism is also directed at the concept of sustainable development in that some consider it to be, in the context of global development, yet another manifestation of Western hegemony. If the process or goal of development is no longer possible in an era of globalization, does this suggest, perhaps, that sustainable development is unachievable? On the one hand, the so-called neoliberal school argues that economic globalization, achieved through free trade and capital mobility, will eventually lead to wealth and prosperity for all. Conversely, the radical school argues that globalization is increasing the rift between the wealthy 'core' and the less developed 'periphery'. What is certain is that the concept of sustainable development is problematic. Not only is it, in a sense, a luxury that many countries suffering from the 'pollution of poverty' simply cannot afford, but it also gives rise to a number of questions, including:

- *What should be developed sustainably?* Personal wealth, national wealth, human society, or ecological diversity?
- *For how long should it be sustained?* A generation, a century, or 'for ever'?

- *Against what baseline can sustainable development be assessed?* No further environmental degradation, or limits of acceptable change?
- *Who is responsible for sustainable development?* Individuals, national governments, or the international community?
- *Under what political-economic conditions is sustainable development possible?*

Answers to these and other questions remain elusive. Nevertheless, although the achievement of sustainable development or, more generally, global environmental sustainability, undoubtedly faces enormous challenges, it is recognized that the alternative (i.e. unsustainability) is not an option.

Sustainable tourism development

As observed in the introduction to this chapter, the roots of sustainable tourism development lie in the strategies for developing alternative forms of tourism that emerged during the late 1980s. As concern grew over the negative consequences of the rapid and uncontrolled growth of mass international tourism, epitomized, perhaps, by the development of the Spanish 'costas' from the 1960s onward, attention was increasingly focused on alternative approaches to tourism development. Variously labelled as 'green', 'responsible', 'appropriate', 'low-impact', 'soft' or 'ecotourism', these styles of tourism collectively represent, literally, an alternative to mass tourism development. Designed to minimize tourism's negative impact while optimizing benefits to the destination, they share a number of characteristics that, collectively, lie in opposition to those of conventional mass tourism (see Table 2.2).

Table 2.2 Characteristics of mass vs alternative tourism

Conventional mass tourism	Alternative forms of tourism
General features	
Rapid development	Slow development
Maximizes	Optimizes
Socially/environmentally inconsiderate	Socially/environmentally considerate
Uncontrolled	Controlled
Short term	Long term
Sectoral	Holistic
Remote control	Local control
Development strategies	
Development without planning	First plan, then develop
Project-led schemes	Concept-led schemes
Tourism development everywhere	Development in suitable places
Concentration on 'honeypots'	Pressures and benefits diffused
New building	Re-use of existing building
Development by outsiders	Local developers
Employees imported	Local employment utilized
Urban architecture	Vernacular architecture

continued

Table 2.2—*continued*

Conventional mass tourism	Alternative forms of tourism
Tourist behaviour	
Large groups	Singles, families, friends
Fixed programme	Spontaneous decisions
Little time	Much time
'Sights'	'Experiences'
Imported lifestyle	Local lifestyle
Comfortable/passive	Demanding/active
Loud	Quiet
Shopping	Bring presents

Source: Adapted from Lane (1990); Butler (1990).

Alternative tourism is considered by some to be synonymous with sustainable tourism and there are, of course, many contemporary examples of such tourism development in practice. Typically, they tend to be small scale and appropriate to the area, with the emphasis on protecting and enhancing the quality of the tourism resource. Ownership and control of tourism development is largely in the hands of local communities and is directed towards optimizing the long-term benefits to visitors, the destinational environment and local people (see Box 2.2 – The Damaraland Camp, Namibia).

Box 2.2

The Damaraland Camp, Namibia

The Damaraland Camp safari lodge is located on the northern banks of the Haub River Valley, some 90 km inland from Namibia's Skeleton Coast in the remote northwest of the country. Comprising eight large walk-in tents, each with en-suite facilities, it provides safari tourists with a luxury base from which to participate in a variety of nature-based activities in the area, including nature drives and walks and visits to the Twyfelfontein engravings. Since it was established in the mid-1990s, the Damaraland Camp has also become widely recognized as one of the leading ecotourism projects in southern Africa. Indeed, reflecting its success in combining tourism development with local community partnerships, environmental conservation and sustainable wildlife, the camp was the winner of the WTTC's Tourism Tomorrow Conservation Award in 2005.

Up until the early 1990s, the Damaraland area was unprotected and open to poachers. No formal conservation measures were in place, wildlife was in decline and unemployment in the local community was virtually 100 per cent. Initially, and with the support of a number of agencies, members of the local community set up a game guard system in order to protect the wildlife from further depletion. Having achieved some success, the community group then considered ways of developing the area sustainably and, consequently, joined forces with Wilderness Safaris in 1996 to establish the Damaraland Safari Camp as a community-based tourism venture. By 1998, the success of the Camp resulted in the area being designated as the Torra Conservancy, one of four Community Wildlife Conservancies in Namibia. The Conservancy is now self-supporting and benefits from a percentage of Wilderness Safaris' income, meeting all its management costs and reinvesting in community projects. At the same time, lodge staff are recruited locally (one local woman who was once a goat herder is now the manager of the lodge), the community benefits economically from tourism and, over the ten years of the project, wildlife populations in the Conservancy have doubled. As a result, the community is thriving, poverty having been alleviated directly through tourism and conservation, and local people having regained, in a sense, ownership of their environment and their future.

Sources: www.namibian.org/travel/lodging/private/damaraland.htm; www.tourismfortomorrow.com.

However, alternative tourism has also been widely criticized for being just that: an alternative, rather than a solution, to the 'problem' of mass tourism. At the same time, the focus on alternative forms of tourism development, particularly on ecotourism (see Plate 2.1) – itself a controversial concept (see Fennell 1999) – has served to amplify the distinction between mass, implicitly 'bad' tourism (and tourists, perhaps), and alternative 'good' forms of tourism. Consequently, although the concept of sustainable tourism development, as originally conceived, applied the principles of sustainable development to tourism more generally, attempts to define and implement sustainable tourism development have, until recently, been undermined by the continuing mass–alternative dichotomy. As one commentator once proposed, for tourism development to be sustainable, it should be based upon 'options or strategies considered preferable to mass tourism' (Pigram 1990).

Plate 2.1 ***South Africa, Dikhololo Resort near Pretoria: Tourists preparing to go on a game-watching outing.***

What is sustainable tourism development?

It is probably true to say that, over the past 15 years, the concept of sustainable tourism development has been the dominant issue in both the study and practice of tourism. Certainly, the subject has spawned innumerable academic books and articles, as well as two dedicated journals (*Journal of Sustainable Tourism* and *Journal of Ecotourism*). Indeed, by 1999, an annotated bibliography of sustainable tourism published by the WTO listed 96 books and 280 articles on the subject (WTO 1999)! Nevertheless, despite the degree of attention paid to it, it has proved difficult, if not impossible, to achieve consensus on a definition of sustainable tourism development. Generally, it has been described as a 'positive approach intended to reduce tensions and friction created by the complex interactions between the tourism industry, visitors, the environment and the communities which are host to holidaymakers' (Bramwell and Lane 1993). However, this reveals neither the developmental objectives of sustainable tourism, nor the processes by which such processes might be achieved and measured.

However, it is logical to suggest that, if sustainable tourism development represents a sector-specific application of the 'parental paradigm' of sustainable development, then it should also share its principles and objectives. In other words, the essential role of tourism development is its contribution to wider economic and social development within the destination. Therefore, sustainable tourism development should be seen simply as a means of achieving sustainable development through tourism. Thus, not only should tourism itself be environmentally sustainable, but it should also contribute indefinitely to broader sustainable development policies and objectives.

Such an approach was, in fact, adopted by the Globe 90 Conference in Canada where, in recognition of tourism's developmental role, three basic principles to guide tourism planning and management were proposed (Cronin 1990):

- Tourism must be a recognized sustainable economic development option, considered equally with other economic activities;
- There must be a relevant tourism information base to permit recognition, analysis and monitoring of the tourism industry in relation to other sectors of the economy;
- Tourism development must be carried out in a way that is compatible with the principles of sustainable development.

This approach has a number of important implications in terms of conceptualizing sustainable tourism development. First, it demands that tourism's developmental potential is assessed against other economic sectors (if the potential to develop other economic activities exists) – that is, tourism is considered within a broader socio-economic and environmental context. Second, this implies that tourism should not compete for scarce resources; rather, emphasis should be placed on the most efficient and sustainable shared use of resources. Third, the mass–sustainable tourism dichotomy referred to earlier becomes irrelevant. In other words, the challenge is to ensure that all forms of tourism (including mass tourism) are planned and managed in such a way as to contribute to sustainable development. Finally, and pointing to a major and as yet unresolved problem, an holistic approach that embraces the entire tourism system is required. That is, while it is important to consider the contribution of tourism to development in the destination, it is no less important to consider the tourism-generating region and, in particular, the environmental consequences of travel to the destination (Høyer 2000).

Subsequently, many tourism planning and policy documents embraced the principles of sustainable development. The typical principles and guidelines in such documents are summarized in Table 2.3.

In particular, the sustainable use of natural resources and the development of tourism within physical and socio-cultural capacities are of fundamental importance, while consideration is given to equitable access to the benefits of tourism. The concept of futurity is also implicit within these guidelines. Moreover, the principle of community involvement appears to satisfy the specific requirements of self-reliance and endogenous development that are critical elements of the sustainable development paradigm.

Table 2.3 Sustainable tourism development: a summary of principles

- The conservation and sustainable use of natural, social and cultural resources is crucial. Therefore, tourism should be planned and managed within environmental limits and with due regard for the long-term appropriate use of natural and human resources.
- Tourism planning, development and operation should be integrated into national and local sustainable development strategies. In particular, consideration should be given to different types of tourism development and the ways in which they link with existing land and resource uses and socio-cultural factors.
- Tourism should support a wide range of local economic activities, taking environmental costs and benefits into account, but it should not be permitted to become an activity which dominates the economic base of an area.
- Local communities should be encouraged and expected to participate in the planning, development and control of tourism with the support of government and the industry. Particular attention should be paid to involving indigenous people, women and minority groups to ensure the equitable distribution of the benefits of tourism.
- All organizations and individuals should respect the culture, economy, way of life, environment and political structures in the destination area.
- All stakeholders within tourism should be educated about the need to develop more sustainable forms of tourism. This includes staff training and raising awareness, through education and marketing tourism responsibly, of sustainability issues among host communities and tourists themselves.
- Research should be undertaken throughout all stages of tourism development and operation to monitor impacts, solve problems and to allow local people and others to respond to changes and take advantage of opportunities.
- All agencies, organizations, businesses and individuals should cooperate and work together to avoid potential conflict and to optimize the benefits to all involved in the development and management of tourism.

Sources: Adapted from: Eber (1992); WTO (1993); ETB (1991); WTO/WTTC (1996); EC (1993).

However, despite this initial adherence to the principles of sustainable development, in practice sustainable tourism development has, more typically, reflected what has been described as a 'tourism centric and parochial' perspective (Hunter 1995). Rather than considering tourism within a broader developmental context, the principal objective has been the development of sustainable tourism or, more precisely, sustaining tourism itself. Consequently, rather than optimizing its contribution to the broader sustainable development of the destination as a whole, attention has been focused primarily on the preservation of the natural, man-made and socio-cultural resource base on which tourism depends in particular settings, thereby permitting the longer term 'survival' of tourism as an economic sector. In short, sustainable tourism development is most commonly taken to represent 'tourism which is in a form which can maintain its viability in an area for an indefinite period of time' (Butler 1993: 29).

To an extent, of course, this is sound business practice. All industries strive to maintain their resource base for longer term survival, while sound environmental policies may significantly enhance profitability. It does not, however, equate with sustainable development. Thus, while there is evidence of many successful sustainable tourism projects, these tend to be localized and small scale; conversely, there is little or no evidence of sustainable tourism development on a wider scale. This, in turn, suggests that 'true' sustainable tourism development (that is, tourism development consistent with the tenets of sustainable development) is unachievable. Indeed, it is now generally accepted that sustainable tourism development, though highly desirable, is an idealistic concept that is difficult, if not impossible, to put into practice.

This is not to say that efforts have not been (and continue to be) made to implement sustainable tourism practices; indeed, as we shall see shortly, many organizations and businesses throughout the tourism system have attempted to adopt or promote the principles of sustainable tourism development. However, it is important to recognize that, as an approach to tourism development, sustainable tourism faces a number of significant challenges.

Sustainable tourism development: weaknesses and challenges

Despite the lack of consensus over the definition and viability of sustainable tourism development, it is probably true to say that, in principle, few would argue against its objectives. Equally, however, few would also argue against the fact that, as a specific economic and social activity, tourism displays a number of characteristics that appear to be in opposition to the concept of sustainable development. These characteristics, referred to in a widely cited paper as tourism's 'fundamental truths' (McKercher 1993), may be summarized as follows:

- As a major, global activity, tourism consumes resources, creates waste and requires significant infrastructural development;
- The development of tourism may, potentially, result in the over-exploitation of resources;

- In order to survive and grow, the tourism industry has to compete for scarce resources;
- The tourism industry is predominantly made up of smaller, private businesses striving for short-term profit maximization;
- As a global, multisectoral industry, tourism is impossible to control;
- 'Tourists are consumers, not anthropologists';
- Most tourists seek relaxation, fun, escape and entertainment; they do not wish to 'work' at being tourists;
- Although an export, tourism experiences are produced and consumed 'on site'.

Collectively, these 'truths' point to three major issues. First, tourism is not a 'smokeless' industry. As has long been recognized, the development of tourism may result in a significant environmental and social impact on destinations, the effective management of which is fundamental to the sustainability of tourism. Second, the consumption of tourism is of direct relevance to its (sustainable) developmental contribution – the scale, scope and nature of the demand for tourism represents a significant challenge to sustainable development. Third, the structure, scale and inherent power relations of the tourism industry raise important questions about the likelihood of a collective, uniform commitment to the principles of sustainable development on the part of the industry.

These issues are discussed shortly and are also considered in more detail in subsequent chapters. However, a further general weakness in the concept of sustainable tourism development deserving of comment is that its principles and objectives have tended to manifest themselves in numerous sets of guidelines that, together, represent a relatively inflexible and arguably Western-centric 'blueprint' for the development of tourism (Southgate and Sharpley 2002). In other words, they propose a relatively uniform approach to tourism development, usually based on managing the limits (according to Western criteria) of acceptable environmental and social change, which is unable to account for the almost infinite diversity of tourism developmental contexts. That is, all destinations differ in terms of a variety of factors, including local developmental needs, local environmental attitudes and knowledge, local governance and planning systems, the maturity and diversity of the local economy, and so on. For example, some less developed countries have, arguably, yet to reach the 'take-off' stage referred to in Chapter 1. As a result, not only are they unable to take advantage of some of the opportunities offered by the development of tourism, but they also face a set of development priorities, such as widespread poverty and malnutrition, that are different to those addressed by sustainable tourism development. Thus, in some countries such as The Gambia (see Box 2.3), an alternative approach to tourism and development may be required.

Interestingly, more recent approaches to sustainable tourism do, in fact, focus on the specific issue of poverty and, in particular, the need to develop policies that spread the benefits of tourism development to the poorest members of destination societies who are unable to establish a formal position in the local tourism system (see WTO 2002; UN 2003). Referred to as pro-poor tourism, this is considered in more detail below.

Box 2.3

Tourism and development in The Gambia

Lying on the west coast of Africa, The Gambia is one of the continent's smallest and poorest countries. It possesses few natural resources and the average annual per capita income among its 1.4 million people, the majority of whom survive on subsistence farming, is just US$278; between 50 and 70 per cent of the population live in extreme poverty. However, the country is blessed with beaches on its Atlantic coastline and virtually unbroken sunshine during the winter months. Consequently, tourism has long made an important contribution to the Gambian economy although, over the past decade, the tourism sector has had little impact on the country's development. Indeed, economic growth has, in recent years, remained stagnant while little progress has been made more generally in overcoming the other development challenges the country faces.

This lack of progress in development reflects, in part, the relatively poor performance of the Gambian tourism industry over the past decade, although this has not always been the case. Following its introduction in 1965, tourism to The Gambia grew steadily, and by the mid-1990s almost 90,000 international arrivals were recorded, the great majority travelling on package tours from, primarily, the UK and Scandinavia. The 1994/5 season, however, witnessed a significant fall in tourism as a direct consequence of the 1994 military coup, and since then tourist arrivals have been erratic. Indeed, as the table below demonstrates, total arrivals in 2004 were almost identical to the 1994 season total of 89,997.

Tourist arrivals in The Gambia 1995–2004

Year	*Arrivals*	*Year*	*Arrivals*
1995–6	72,098	2000	78,710
1996	76,814	2001	75,209
1997	84,751	2002	78,893
1998	91,106	2003	73,000
1999	96,122	2004	90,098

This decline in tourism cannot be explained by a lack of development of the sector within The Gambia – since 1995 significant investment has been made in accommodation facilities, the road network and airport. Nevertheless, as a destination The Gambia faces a number of problems, including:

- High dependence on a small number of key markets;
- Limited appeal to an equally limited winter-sun market;
- Few opportunities for diversification of the tourism product;
- Dependence on overseas tour operators;
- A continuing reputation for 'hassle' by local youths.

At the same time, however, tourism has consistently failed to meet expectations as an engine of economic growth and development in The Gambia. Although the industry accounts for some 12 per cent of GDP and up to 20 per cent of formal employment, for a number of reasons the country has been unable to take advantage of the opportunities tourism offers. The domination of the sector by overseas tour operators means that the local tourism authorities have little control over development, while a significant proportion of the profits from tourism are repatriated. Equally, virtually all the goods required by the tourism industry have to be imported, and the local economy lacks the skills, diversity and ability to produce and supply goods in sufficient quantity or of appropriate quality. Most importantly, however, The Gambia lacks the financial and human resources to develop tourism effectively and, although relatively large sums of money are generated through sales taxes and the tourist arrivals tax, little is reinvested in the tourism industry or potential supporting sectors, such as agriculture. In short, the country suffers from economic and political immaturity that prevents the realization of tourism's developmental potential.

Source: Sharpley (2006a).

Tourism as sustainable development?

In addition to the general concerns about the validity of the concept of sustainable tourism development discussed above, there are a number of specific areas in which tourism, as a particular vehicle of development, is unable to meet the principles and required processes of sustainable development. These reflect some of the 'truths' of tourism and are considered in detail elsewhere (Sharpley 2000). Nevertheless, a brief review of these points of divergence between sustainable tourism development and the broader sustainable development paradigm is useful, for two reasons:

1. It goes some way towards explaining the inevitability of the tourism-centric, localized and small-scale focus of most sustainable tourism projects in practice.
2. It points to ways in which tourism may, conversely, make a contribution to the sustainable development of destination areas.

With reference to Table 2.1 above, the development of tourism is unable to meet sustainable development's fundamental principles, nor its development and sustainability objectives:

- *Holistic approach*. All tourism development should be considered within a global socio-economic, political and ecological context. More simply stated, all elements of the tourism experience should be sustainable. However, given the breadth of the tourism system and the fragmented, multisectoral character of the tourism industry, such an approach is difficult, if not impossible. Within this context, a particular issue is the unsustainability of most modes of transport (Høyer 2000). By definition, tourism involves travel by land, sea or air. The environmental impact of air travel (both infrastructural developments and aircraft emissions) has emerged as a particular area of contemporary concern. Becken and Simmons (2005) suggest that tourism's role in the context of greenhouse gas emissions will increase, especially given the fact that the industry is growing worldwide and that long-distance travel is increasing in popularity. As a result of this concern, some organizations have introduced innovative schemes. For example, in 2005 British Airways announced a scheme whereby its passengers could offset the environmental cost of their journey by making voluntary donations equivalent to that environmental cost, which help to invest in projects seeking to reduce global carbon dioxide levels. A second example is the Carbon Neutral Company which has an Internet site where, after travellers calculate their CO_2 emissions, they can contribute to carbon offset projects, several of which are located in developing countries and include forestry projects. However, such carbon offset initiatives remain voluntary and have little effect on worldwide emissions related to transport and travel.

- *Futurity*. The tourism industry is not only diverse and fragmented, it also largely comprises small, private sector, profit-motivated businesses. Therefore, although there is undoubtedly evidence of tour operators and other organizations who have adopted a 'responsible', longer term perspective (Mann 2000; www.responsibletravel.com), it is likely that most businesses are more concerned with short-term profit than the long-term sustainable development of a destination.
- *Equity*. Although tourism development by no means leads inevitably to destinational dependency, the structure, ownership and control of the tourism industry on a global scale, as well as the regionalized and polarized nature of international tourist flows, suggests that inter- and intra-generational equity is unlikely to be achieved through tourism. In other words, although there are many examples of community-based tourism projects, the overall political economy of the tourism system is such that tourist flows and the tourism industry itself are generally dominated by Western-owned global networks (Brohman 1996b). Moreover, the tourism system within destinations or countries also tends to be dominated by the local élite, restricting equitable access to the benefits of tourism.
- *Development objectives*. Tourism has, of course, long been considered an effective means of achieving development. However, 'development' in the tourism context usually means traditional economic development rather than the more contemporary interpretations of its objectives. That is, tourism undoubtedly represents an important source of income, foreign exchange and employment. However, it remains unclear to what extent broader (sustainable) developmental goals, such as the satisfaction of basic needs, self-reliance and endogenous development, can be achieved through tourism. Moreover, not only are the benefits of tourism usually restricted to particular geographical areas and/or sectors of local destinational communities – as noted below, so-called 'pro-poor tourism' attempts to meet this challenge – but also the very nature of tourism as a discretionary and fragile form of consumption is such that the ability of destinations to manage tourism development may be compromised by factors beyond their control. Thus it is increasingly recognized that while tourism contributes to economic growth, it does not necessarily lead to 'development'.
- *Sustainability objectives*. A fundamental principle of all sustainable tourism development policies is that the natural, social and cultural resources upon which tourism depends should be protected and enhanced. Furthermore

> most, if not all, sectors of the tourism industry have a vested interest in following such a policy. This may result from either a genuine commitment to sound environmental practice or from more pragmatic business reasons. Either way, the extent to which sustainability objectives are achievable remains questionable. Resource sustainability is dependent on all sectors involved directly and indirectly in the tourism industry working towards common goals. Although different organizations and industry sectors have, to a lesser or greater extent, adopted environmental management policies (such as British Airways' carbon offset initiative mentioned above), sustainability in tourism will only be achieved when the industry as a whole accepts the need for such policies.

Most significantly, the achievement of sustainable tourism development is dependent on a number of prerequisites, in particular the adoption of a new 'social paradigm' regarding the consumption of tourism (or, more simply stated, the need for all tourists to become 'good' or 'responsible'), and the emergence of global political and economic systems dedicated to more equitable resource use and development. Both these issues are considered in more detail later in this book although, as this chapter goes on to conclude, perhaps the greatest challenge facing progress towards the achievement of sustainable tourism development (and, indeed, sustainable development more generally) are the global political-economic structures and trend towards globalization.

However, it is important to note that, despite the undoubted difficulties in implementing 'true' sustainable tourism development, various approaches have been adopted by businesses and organizations within the tourism system, from accreditation schemes and 'eco-labelling' to more direct intervention in promoting sustainable tourism development. These are by no means universal; for example, Forsyth (1995) found little evidence of widespread adherence to sustainable business and development principles within the UK travel and tourism industry, although a number of new initiatives have evolved over the past decade. Nevertheless, they serve to demonstrate a commitment to the objectives of sustainable tourism development within different levels and sectors of the tourism system.

Sustainable tourism development in practice

A complete review of sustainable tourism development in practice is well beyond the scope of this chapter. Innumerable examples exist of codes of conduct, tourism development guidelines, specific sustainable/ecotourism projects, industry sector initiatives, NGO activities and so on, while some projects or initiatives enjoy wider recognition than others. For example, Zimbabwe's CAMPFIRE programme (Communal Areas Management Programme for Indigenous Resources), established in the late 1980s and now involving some 30 communities throughout the country, is well known as a pioneering and successful project that has promoted wildlife conservation and community development through tourism. Conversely, there are many smaller scale local initiatives that have not attracted such wide attention; for example, the Tumani Tenda ecotourism camp in The Gambia, established in 1999 and supporting a village community of 300, is virtually unknown outside that country (Jones 2005).

Moreover, the promotion or implementation of sustainable tourism development occurs at different levels within the tourism system (international, national, local), in different sectors (private, public and voluntary), and in a number of different ways. At the same time, various activities and schemes may also be interrelated. For example, the International Federation of Tour Operators (IFTO) established the *Tourism for Tomorrow* awards in 1989 in order to encourage the tourism industry as a whole to adopt environmentally sound practices. Sponsorship was taken over by British Airways in 1992 to extend global awareness of the awards and, since 2004, they have been run under the auspices of the WTTC (www.tourismfortomorrow.com). The aim of these awards is to promote best (sustainability) practice across all sectors of the global tourism industry, based on the policies for sustainable tourism development outlined in the WTTC's *Blueprint for New Tourism* (WTTC 2003). Thus, it is difficult to establish a full and clear picture of how sustainable tourism development is manifested in practice.

Nevertheless, for the purposes of this chapter, it is useful to explore the application of the principles of sustainable tourism development from the perspective of different policies/initiatives and, where relevant, different sectors of the tourism system. Many specific examples are, of course, provided in the tourism literature, while a recent document published jointly by the United Nations Environment Programme and the WTO (UNEP/WTO 2005) provides a number of contemporary case studies together with a comprehensive guide to official documentation and other information sources relating to sustainable tourism development. The principal areas of activity are discussed below.

Sustainable tourism development guidelines

Perhaps the most common way in which the principles of sustainable tourism development are promoted is via the publication of guidelines for development or through actual policy documents. Guidelines or principles for sustainable tourism development have been published at national, regional and international levels by both the public and voluntary sectors. One of the earliest set of guidelines, for example, was published by the Wordwide Fund for Nature (WWF) in collaboration with the pressure group Tourism Concern (Eber 1992), while the WTO, WTTC and other global organizations have produced guidelines, statements of good practice and other documents extolling the virtues of sustainable tourism and, more specifically, ecotourism. Inevitably, of course, such documents are simply guidelines; there is no requirement, legal or otherwise, for businesses or destinations to adopt them. Conversely, a number of countries, such as Costa Rica, have produced tourism development policies that explicitly embrace the principles of sustainable development and, in particular, ecotourism. One of the first, and better known, was Australia's National Ecotourism strategy, published in 1994, subsequently followed in 2003 by a Tourism White Paper which embraced sustainable tourism more broadly. Importantly, the National Ecotourism strategy was supported by grant funding totalling AUS$10 million, ensuring that its policies would be actioned. Guidelines on sustainable tourism have also been produced for particular types of destination (for example, national parks) or particular forms of tourism, such as rural tourism (Countryside Commission 1995).

One of the most recently published documents, at the time of writing, is *Making Tourism More Sustainable: a guide for policy makers* (UNEP/WTO 2005). This is significant in two respects. First, in recognition of the fact that it is governments that develop and implement policy, it is aimed specifically at regional or national tourism policy makers. Second, it signifies a departure from the rigid 'blueprint' approach to sustainable tourism development by stating that 'it must be clear that the term "sustainable tourism" – meaning "tourism that is based on the principles of sustainable development" – refers to a fundamental objective: to make all tourism more sustainable' (2005: 11). It goes on to identify 12 aims that represent an agenda for sustainable tourism (Table 2.4). However, the extent to which such an agenda, closely reflecting the objectives of sustainable development, is achievable remains questionable.

Table 2.4 Agenda for sustainable tourism

1 Economic viability	5 Visitor fulfilment	9 Physical integrity
2 Local prosperity	6 Local control	10 Biological diversity
3 Employment quality	7 Community well-being	11 Resource efficiency
4 Social equity	8 Cultural richness	12 Environmental purity

Source: Adapted from UNEP/WTO (2005).

Accreditation schemes

A popular means of promoting sustainable practice within the tourism industry is through voluntary accreditation schemes and, more specifically, eco-labelling or other certification schemes which identify products or services that meet specific environmental or quality benchmarks. Many such schemes exist, particularly at the national level; the only international accreditation scheme is Green Globe 21. Originally launched as Green Globe by the WTTC in 1994 as a membership- and commitment-based programme (www.greenglobe21.com), the scheme has evolved into a benchmarking and certification system for sustainable tourism. Environmental improvements are monitored through annual benchmarking, and the organization also provides technical support in a number of countries. At the national level, funding was provided in Australia under the National Ecotourism strategy mentioned above to support the development of the National Ecotourism Accreditation Programme (NEAP), now called the EcoCertification Programme. This certifies the environmental credentials of products (e.g. tours, hotels) rather than the businesses providing them. Although there is evidence to suggest that such accreditation schemes encourage higher environmental standards in the tourism industry, the overall benefit of certification or eco-labelling schemes remains the subject of debate.

Codes of conduct

Codes of conduct have long been used as a means of trying to influence the behaviour or practices of individuals and organizations without recourse to laws and regulations (see Plate 2.2). In tourism, such codes are numerous and are aimed at tourism businesses, visitors and policy makers, and are formulated by international agencies, governments, NGO/voluntary organizations or even bodies representing particular sectors of the industry as a means of self-regulation (Mason and Mowforth 1995). Pattulo with Minelli (2006), for example, has published the *Ethical Travel Guide* for Tourism Concern. Codes of practice are utilized where it is inappropriate or difficult to impose regulations; however, they rely on voluntary action and there is, of course, no mechanism for monitoring or ensuring compliance. Nevertheless, they are widely used as a means of trying to encourage tourists to behave more responsibly, although some see this as evidence of the creeping 'moralization' of tourism (Butcher 2002).

Plate 2.2 ***Russia, St Petersburg: Codes of conduct for tourists visiting the Peter and Paul Fortress.***

Industry initiatives

Sustainable tourism development initiatives are, arguably, most effective when implemented by the tourism industry itself. In other words, Corporate Social Responsibility (CSR) is considered to be a key element in the achievement of sustainable development in general. For the tourism industry, this means adopting environmental and responsible practices in the day-to-day running of the business and, in the case of tour operations and other international activities, addressing sustainability issues at the destination. An enormous variety of initiatives are in evidence in the tourism industry, from individual organizations or tour operators offering holidays designed to involve tourists in local community or environmental projects to local, community-run projects (Mann 2000). There are also a number of schemes that involve groups of businesses in working towards sustainable tourism development, an early example being the ECOMOST project, established by the International Federation of Tour Operators, which sought to increase the sustainability of tourism in the Balearics (IFTO 1994). Contemporary initiatives include the following:

- *The International Tourism Partnership (ITP)*. This has developed from the successful International Hotels Environment initiative originally launched in 1992 and is now a programme of the Prince of Wales International Business Leaders Forum. Within ITP, airlines, hotels, tour operators and NGOs work together to promote sustainable business practice within travel and tourism. The BEST scheme (Business Enterprises for Sustainable Travel) also operates as a programme of the ITP, its objective being to develop and support innovative travel and tourism practices that promote sustainability among businesses, local communities and travellers themselves.
- *Tour Operators' Initiative (TOI)*. Founded in 2000, the TOI is a voluntary, non-profit initiative open to all tour operators. It has been developed by tour operators for tour operators with the support of UNEP, UNESCO and the WTO, who are also full members of the initiative. Its purpose is to develop and implement sustainable tourism development practices among the sector, with a focus on four particular areas of activity: sustainability reporting; cooperation with destinations; supply chain management; and communication.

Other initiatives include the Travel Foundation (www.thetravelfoundation.org), a UK-based charity that works with outbound tour operators to manage tourism more sustainably in the destination. It does so by developing projects that boost the benefits of tourism to destination communities, promote the conservation of the local environment and culture, and enhance tourists' experiences. There also exist a number of organizations that market 'responsible' holidays; for example, Responsibletravel.com is an online travel agent that provides holidays 'for people who've had enough of mass tourism' (www.responsibletravel.com).

Voluntary sector/NGOs

The concept of sustainable tourism development and, indeed, efforts to promote it have long been underpinned by a number of voluntary organizations and pressure groups. Since 1989, the London-based Tourism Concern has campaigned to raise awareness of tourism's negative consequences and worked with the industry and destinations to enhance tourism's developmental contribution. Similar groups operate in other countries, such as Studienkreis fur Tourismus und Entwicklung in Germany, while research has revealed the extent to which grass-roots movements in some countries oppose inappropriate tourism developments (Kousis 2000). At the international level, Green Globe, though strictly a travel industry-based organization, is also an NGO that promotes responsible practices throughout the sector.

Pro-poor tourism

One of the most recent approaches to tourism development is so-called pro-poor tourism (Ashley *et al.* 2000). Given its specific focus on poverty reduction, pro-poor tourism does not necessarily reflect the broader developmental aims of sustainable development. However, the WTO has attempted to link the two in its Sustainable Tourism–Eliminating Poverty (ST–EP) programme. Originating in 2003, the programme is tied to the UN Millennium Development Goals as laid out in Box 1.1, and its international office is located in Seoul. Pro-poor tourism initiatives seek to achieve greater equity by providing the poorest members of destination societies with the opportunity to benefit from access to tourism markets (see Chapter 5). In other words, the poor are frequently excluded from the local tourism sector, and thus are unable to sell locally produced products or provide other services. The objective of pro-poor tourism, therefore, is to open up access for the poor to the tourism sector, thereby providing them with a vital source of income. There are many examples of successful pro-poor tourism in practice (www.propoortourism.org.uk), although the need to specifically target the poor in tourism development points to the failure of previous development policies to spread the benefits of tourism equitably.

There is no doubt, then, that attempts to apply the principles of sustainable development are occurring in different ways and in different sectors across the tourism system. There is also no doubt that, at the level of individual initiatives or projects, significant progress has been made in achieving greater sustainability in tourism development. However, tourism is a global activity; only when the majority of travel and tourism businesses and organizations (and, indeed, tourists themselves) around the world adopt the principles of sustainability will sustainable tourism development become a realistic objective. Therefore, the final question to be addressed is: What are the implications of transformations in the global political-economy, and globalization in particular, for sustainable tourism development?

Sustainable tourism and globalization

In recent years, increasing attention has been paid by academics to the relationship between tourism and globalization, both generally (Knowles *et al.* 2001; Wahab and Cooper 2001) and in the specific context of tourism and development (Bianchi 2002). With reductions in trade barriers, the tourism market is becoming more globalized. There is pressure for countries to open up their borders if they want to take part in the global economy. Multinationals have the ability to operate across borders with fewer restrictions. As such, the multinational corporations hold a great deal of power in where they choose to locate, how they operate and where they get the resources they need. This raises some very profound issues in terms of the three corners of sustainability (economic, environmental and social). Reid (2003: 3) cites the work of Van den Bor *et al.* (1997), who state that in the new global economy capital can locate 'wherever the costs of production are the lowest and where social and environmental restrictions are fewest'. In some regards there has been a shift in power from the state to the corporation. Governments in developing countries are forced to compete against each other in offering various incentives to try to attract multinationals, as these companies bring with them name recognition and access to tourist markets that developing countries need. Bianchi (2002) addresses the central concern of the political economy of tourism as not merely whether incomes are rising from tourism, or whether the large multinationals provide a decent wage, but rather the extent to which different modalities of global tourism are leading to an increase or reduction in inequality of access to power and resources. He also raises the question as to whether new areas of discussion, such as fair trade and related development programmes explored in Chapter 5, represent isolated examples of endogenous development, or begin to challenge the very asymmetrical structures of tourism production and exchange. The following chapter will explore the complex relationship between tourism and globalization in greater detail.

Discussion questions

1. **Why is there so much controversy surrounding sustainable tourism development?**
2. **Does sustainable development represent a form of Western imperialism?**
3. **Are new, evolving, alternative forms of tourism development more sustainable?**
4. **How can a specific tourism development or a tourism industry/organization be made more sustainable?**
5. **What roles do governments, NGOs, industries and individuals play in the implementation of sustainable tourism development?**

Further reading

Baker, S. (2006) *Sustainable Development*. London: Routledge. [Contains a good overview of sustainable development including the challenges of promoting sustainable development in different social, political and economic contexts.]

UNEP/WTO (2005) *Making Tourism More Sustainable: A Guide for Policy Makers*, Paris/Madrid: United Nations Environment Programme/World Tourism Organization. [This document (which can be downloaded from the UNEP website: www.unep.org) provides a comprehensive and contemporary review of appropriate policies and processes for the achievement of sustainability through tourism development. A number of excellent, in-depth case studies of sustainable tourism 'in action' supplement the main text.]

Websites

Part of the International Business Leaders' Forum, this is a partnership of leading travel and tourism businesses. It provides the industry with the knowledge and ability to achieve more responsible practices in travel and tourism: www.internationaltourismpartnership.org.

The website of Business Enterprises for Sustainable Tourism (BEST), an organization that seeks to be a leading source of knowledge on travel and tourism practices that advance sustainable development through tourism: www.sustainabletravel.org.

The Tour Operators' Initiative, an international network of tour operators, supported by the UNWTO, that develops and promotes means of minimizing tourism's negative impact and encouraging sustainable development: www.toinitiative.org.

Details of the annual Tourism for Tomorrow Awards: www.tourismfortomorrow.com.

3 Globalization and tourism

Learning objectives

When you have finished reading this chapter, you should be able to:

- **Understand the influences of globalization on tourism development;**
- **Identify the power of multinational tourism corporations in tourism development;**
- **Be familiar with the economic, political and cultural aspects of tourism and globalization;**
- **Be aware of the political economy of tourism and the links to globalization.**

Tourism is very much an agent of globalization with multinational tourism corporations providing services to people as they move across the globe, bringing with them money, values and patterns of consumption. Globalization represents greater integration at a global level in a wide variety of areas, such as trade, finance, communication, information and culture. Facilitated by improvements in transportation and information technology, it has led to time–space compression where people, goods and information travel greater distances and cross political borders in shorter periods of time. Globalization comprises 'trans-state' processes that operate not merely across borders but as if the borders are not there (Taylor *et al.* 2002). Corporations conduct business at international scales presenting challenges for nation states (Reid 2003), and these global interactions have impacts in terms of economies, cultures, politics and environments. Globalization is a much-contested term with both supporters and critics. Some argue that trade liberalization and open borders promotes economic growth, while others argue that the gap between rich and poor is increasing and globalization results in a loss of national identity, culture and control. The tourism development dilemma means that countries interested in pursuing tourism as an agent of development must enter a very competitive global market where the processes of globalization are unevenly distributed, complex and volatile (Momsen 2004).

Globalization and tourism interact in a variety of dimensions, as revealed in the following example of a tourist on a mass tourism holiday package in a developing country. A key element to keep in mind is the power and control issues that accompany the globalization process associated with tourism. A tourist departing on a low-cost airline or charter flight for a two-week holiday at an all-inclusive beach resort in a developing country will likely have made the booking through a travel agent selling holiday packages for a multinational tour operator based in a developed country, or will have self-booked through the Internet. Both the travel agent and the individual tourist will have used the Internet to connect them to evolving and dynamic global distribution channels. The tour company selected by the tourist will typically be vertically integrated, owning or having strategic alliances with airlines, travel agents and hotels, and little of what the individual has paid for the holiday in their home country will go to the destination. The accommodation will probably be reserved in a multinational hotel company with a recognizable brand name which, in the tourist's mind, offers a standardized product with the services and food they are used to while at home. At the hotel the tourist will be able to surf the Internet and watch global news coverage on channels such as BBC World Service or CNN International. Shopping and eating away from the hotel may well be at internationally recognized brand name stores and restaurants. While in the destination the tourist will pay for most things with one of the credit cards issued by the few multinationals that control global personal finance (Sklair 1995). With global banking, the hotel will be able to transfer profits out of the developing country and, with access to international courier services, the executive chef at the hotel, who is probably a citizen of a developed country, can import products from any part of the globe.

In such an unequal power relationship, developing countries are often at the mercy of multinational corporations to market their destination, construct and/or manage hotels and fly in the tourists, and hence the political economy of tourism. While the globalization of the tourism industry may have many advantages, facilitating the arrival of money-carrying guests, there are also many concerns as the local is quickly brought into contact with the global. Conflicts result from clashes in cultures, religion and family values as shifts occur in the structure of the labour force (Wall and Mathieson 2006). It has also been argued that gender inequalities are reinforced where women are often portrayed as either domestic help or the desirable exotic 'other' as part of the sex tourism industry (Momsen 2004). These impacts will be discussed in greater detail in Chapter 7. The power of tour companies lies in their ability simply to change destinations. If a given hotel in a destination does not keep up with the demands of the tourists and/or the tour operator, the tour operator may in fact choose to take their business elsewhere, leaving the hotel scrambling to fill vacant rooms. Tourism is a very competitive industry and tour operators need to be able to make a profit. While tourism is clearly not responsible for all the negative effects of globalization it is certainly part of the process. Part of the development dilemma within the framework of sustainability is how the industry can operate so that the destination developmental needs are taken into account and how the power that multinationals hold can be used to that end.

The purpose of this chapter is to outline the relationship between tourism and globalization. It will first explore the nature of globalization and identify some of the processes facilitating greater global interconnectedness. As Hall (2005) suggests, globalization is a far-reaching idea that embraces nearly all aspects of contemporary culture, identity, governance and economy. Globalization is highly complex and controversial, with some arguing that benefits accrue mainly to multinationals in developed countries. However, in many cases there is limited choice for developing countries other than to participate in the global economy and many governments welcome the multinationals. Allen (1995) suggests that there are three broad strands to examining globalization – economic, political and social – and the chapter will address the relationship between tourism and globalization in each of these three areas.

Processes of globalization

Globalization is about increasing mobility across frontiers in terms of goods and commodities, information and communication, products and services, and people (Robins 1997). The process of globalization has been facilitated by a number of trends including increased market liberalization and trade, foreign investment, privatization, financial deregulation, rapid technological change, automation, changes in transportation and communications, standardization and immigration (Bianchi 2002; Momsen 2004; Weinstein 2005). Scholte (2005) highlights rationalism, capitalism, technological innovation and regulation that facilitate globalization as the four primary forces that have generated the emergence and expansion of supra-territorial spaces. Harvey (1989) introduced the concept of space–time compression whereby advances in technology, transportation and communication have sped up the pace of life, overcoming spatial barriers where the world seems to collapse inwards upon us. In terms of transportation, Airbus has recently tested the Airbus A380, a double-decker airline with a capacity of as many as 555 seats. The rise of the low-cost or no-frills airlines, such as EasyJet and Ryanair in Europe, has been very successful in opening up secondary airports operating on short-haul flights. EasyJet, for example, is opening up routes to Eastern Europe as well as to Marrakech, Morocco in northern Africa. In Canada, WestJet flies to Nassau, Bahamas. Keys to the success of low-cost airlines have been low fares, use of Internet booking (90 per cent of their business), short turnaround times, high plane usage and paperless ticket distribution (Leinbach and Bowen 2004). There has been a rapid growth in Internet sites selling discount airline tickets, hotel rooms and rental cars, thereby reinventing and expanding the concept of distributed data which allowed computer reservations systems to revolutionize air travel in the 1980s (Leinbach and Bowen 2004). The 24-hour global news media processes and packages news events from around the world and delivers them over IP (Internet Protocol) networks, cable, fibre and through satellite and cell phone transmissions to viewers (Rain and Brooker-Gross 2004). In noting the complexity of technology, Wilbanks (2004: 7) suggests that technology 'underlines many structures for exercising power and control from the expanding reach of globalisation to new opportunities for local empowerment'. In this media age, we are also constantly surrounded by commercial and informational messages in our homes, workplaces and landscapes, from product placement in movies, to outdoor advertising to network news packaged for long-distance jet passengers (Rain and Brooker-Gross 2004). The information we receive through advances in technology not only makes us aware of new destinations and travel products but also of natural disasters, war, health concerns and terrorist attacks turning tourists away from destinations. For example, in October 2005 when Hurricane Wilma hit the Mexican resort area of Cancún, images were rapidly sent out over the Internet and through news services, alerting potential tourists of the situation. These brief examples in tourism reveal the nature of the increasing connectivity in our world.

Within the processes of globalization there has been a transformation in the means of production and consumption. It has been suggested that there has been a shift in industrial production systems from Fordism to post-Fordism, also known as flexible specialization, starting in the 1970s (Harvey 1989). Fordism is focused on economies of scale characterized by mass production of standardized goods. In the context of tourism, Fordism is evident in mass standardized package tours where the tourism industry determines the quality and type of product typically sold to mass tourists (Ioannides and Debbage 1998; Mowforth and Munt 2003). While evidence of Fordism is still strongly identifiable (and in many destinations the dominant form of development) in the tourism industry, there is evidence that some parts of the industry are adopting more flexible production systems characteristic of the trend in post-Fordism. The focus in post-Fordism is on economies of scope or network-based economies and high levels of product differentiation through focusing on small batch production of specialized commodities targeting niche markets (Ioannides and Debbage 1998). In tourism there has been a trend towards specialized operators offering tailor-made holidays with niche marketing to more independent or experienced tourists. There is reliance to a greater degree on information technology, integration, networks and strategic alliances. As production has changed, so has consumption. Urry (1990) outlines in some detail the characteristics of post-Fordism consumption where consumption dominates production:

- There are new forms of credit for consumers allowing higher levels of purchasing and debt;
- Most aspects of life are being commodified;
- There is greater differentiation of purchasing patterns by different market segments;
- There is greater volatility of consumer preferences;
- There is consumer reaction against being part of a 'mass' and so there is a need for producers to be consumer driven (Urry 1990).

These shifts in consumption have been characterized by Poon (1989) as changes from 'old tourism' which is typified by the standardized mass package holiday, to 'new tourism' which is based on flexible and customized holidays targeted at specific segments who are often more experienced travellers. These changes in production and consumption have taken tourists to the far corners of the globe as some tourists seek to stay away from the mass tourism destinations. Several authors (Harvey 1989; Urry 1990; Uriely 1997; Mowforth and Munt 2003) have linked the changes outlined above to the shift from modernism to postmodernism which involves dissolving the 'boundaries, not only between high and low cultures, but also between different cultural forms such as tourism, art, education, photography, television, music, sport, shopping and architecture' (Urry 1990: 83). It is about breaking down barriers and accepting other ways or different perspectives, and so the mass tourism package is shifting to customization. Uriely (1997) suggests that trends such as the small and specialized travel agencies, the growing attraction of nostalgia and heritage tourism along with the increase in nature-oriented tourism and simulated tourism-related environments are labelled aspects of postmodern tourism. As the means of production and consumption (see Chapter 6) are changing to be more flexible and customized, tour companies are outsourcing components of the production process to different companies which are sometimes located in different countries. The shift in the production process is not necessarily uniform in destination, as Torres (2002) discovered in Cancún where she found evidence of Fordist and post-Fordist elements. These elements were manifest in different 'shades' of mass tourism, 'neo-Fordism' and 'mass customization'. While these elements may represent examples of the complex functional aspects of globalization, there are also the realms of policy, politics and trade which are also key elements in facilitating globalization.

Globalization is not new; throughout history there have been waves of global economic integration that have risen and fallen (Dollar 2005). In the recent wave of economic integration there has been a dramatic shift in the nature of international trade, from exports of primary products to exports of manufactured products and, along with this, service exports including tourism and software have increased enormously (Dollar 2005). One of the key elements of globalization is the effect of market liberalization opening up national economies to the global market (McMichael 2004). The shift towards market liberalization is evident in the comment by Dollar (2005) that increased levels of integration have been facilitated partly through deliberate policy changes, where the majority of the developing world (measured by population) has moved from an inward- to a more outward-focused strategy, reflected in huge increases in trade integration. In fact, in some cases corporations in developing countries are expanding their operations beyond their borders, referred to as globalization's offspring (*The Economist* 2007). In the context of tourism, the Russian airline Aeroflot is a major bidder for a controlling stake in Alitalia, Italy's national airline (*The Economist* 2007). Some low-cost airlines are also about to expand into neighbouring states. Thailand-based One-Two-Go, for example, is about to offer flights to Cambodia, Singapore, Malaysia, Bali and Bangladesh (Global Travel Industry News 2007). McMichael (2004) suggests that liberalization involves downgrading social goals of national development along with upgrading participation in the world markets through policies such as tariff reduction, export promotion, financial deregulation and relaxation of foreign investment.

Influence of the Bretton Woods meetings

In addition to changes in the economies of countries seeking to participate willingly or otherwise in the global market, there has been a rise in global financial institutions governing world trade and providing loans to countries, which are also driving market liberalization and globalization (McMichael 2004). A number of key institutions and regulatory frameworks were established out of the Bretton Woods meetings following the Second World War. Sometimes referred to as the Bretton Woods institutions (named after the ski resort in New Hampshire, USA, where representatives of 45 countries met from 1 to 22 July 1944), these institutions include the International Monetary Fund (IMF), and the International Bank for Reconstruction and Development (IBRD), which was one of the institutions of the World Bank Group. Subsequent meetings led to the signing of the General Agreement on Tariffs and Trade (GATT) in Geneva in 1947. GATT was a multilateral treaty negotiated in 'rounds' to promote trade and ward off protectionism (Roberts 2002). The World Trade Organisation was formed in January 1995 out of the GATT Uruguay Round. The World Trade Organisation is often seen as the successor of GATT and in 2005 there were 149 member countries. The World Trade Organisation is different from a treaty, as it has independent jurisdiction and the ability to enforce its rules on member states. The World Trade Organisation can set rules on the movement of goods, money and productive facilities across borders, thereby restricting countries in passing legislation or policies that discriminate such movement (McMichael 2004). The World Trade Organisation releases Ministerial Declarations following meetings, and the Declaration from Doha in 2001 stated that 'International trade can play a major role in the promotion of economic development and the alleviation of poverty. We recognize the need for all our peoples to benefit from the increased opportunities and welfare gains that the multilateral trading system generates' (World Trade Organisation 2001). The World Trade Organisation also administers the General Agreement on Trade in Services (GATS), which includes tourism under the category of 'Tourism and Travel Related Services'. The aim of GATS is to liberalize trade in services so that member countries have to allow foreign-owned companies free access to their markets and with no favouritism towards domestic companies (Scheyvens 2002). This would give foreign-owned companies with greater economic power an advantage over small domestic companies.

Globalization and structural adjustment

The shift towards greater market liberalization has also been part of the conditions applied to many of the loans to developing countries in the Structural Adjustment Programmes (SAPs) issued by international lending agencies such as the World Bank and the International Monetary Fund in the 1980s and 1990s. Conditions of these loans involved receiving governments adjusting their economic structures and political policies to be more open to global trade as well as reducing the size and level of involvement of the government in the economy. A more recent trend in these loans has been the inclusion of the condition of state effectiveness and good governance (McMichael 2004), along with a focus on poverty reduction. In 1999 the IMF and the World Bank introduced Poverty Reduction Strategy Papers (PRSPs) as a follow-up to the SAPs (Mowforth and Munt 2003). The PRSPs are a participatory process involving local governments and donor agencies to map out the 'macroeconomic, structural and social policies and programs that a country will pursue over several years to promote broad-based growth and reduce poverty, as well as external financing needs and the associated sources of financing' (IMF 2006). The aim of the programme is to bring development more into line with the United Nations Millennium Development Goals (see Chapter 1). Tourism has been adopted in a number of developing country PRSPs, including those not readily associated with tourism such as Bangladesh, the Central African Republic, Sierra Leone and Guinea (PPT 2004) (See Chapter 4). Mowforth and Munt (2003) note, however, that critics claim that PRSPs are really another form of SAPs with language about poverty reduction. These programmes have been criticized for the reduction in the power of the state, including the reduction of state social welfare programmes.

Economic trading blocs

Economic integration can also be seen at a more regional level in the form of free trade zones and economic unions or partnerships. Examples include the North American Free Trade Agreement (NAFTA) and the European Union. Both these examples include the integration of developing or transitional economies with developed economies. Other examples include the Association of Southeast Asian Nations (ASEAN), Mercosur (South American trading bloc), the Southern African Development Community (SADCC), and the Singapore, Indonesia and Malaysia Growth Triangle. Within this Growth Triangle the resort of Bintan, Indonesia (a 55-minute ferry ride from Singapore) offers seven hotels, four golf courses and five spas (Bintan Resorts 2006). This increased shift towards integrated economies suggests that the economic and social well-being of countries, regions and cities everywhere depends increasingly on complex interactions where systems of production, trade and consumption have become global in scale and scope (Knox *et al.* 2003). The following section will address the economic aspects of tourism and globalization.

Economic aspects of tourism and globalization

Hoogvelt (1997) suggests that there are three key features to economic globalization and these are evident in tourism. The first is that there is a global market discipline whereby, in the face of global competition, individuals, groups and national governments have to conform to international standards of price and quality. The second feature is flexible accumulation through global webs. Companies are organizing through networks or global webs where production activities and services are brought in (often managed electronically) for the short term and, rather than owning suppliers, they are treated as independent agents; however, control is still maintained by the parent company. As an example, services and/or production are contracted out across the globe where production costs (including labour) are cheapest. The system of contracting out externalizes risk as far as possible for the parent company (Bianchi 2002). The third key feature is financial deepening which relates to the fact that money is quickly moved across borders and that profits are increasingly being made out of the circulation of money. As indicated by Scholte (2000), the movement of money has been facilitated by regulation. Scholte (2000: 105) states that regulation has promoted globalization in the following ways:

- Technical and procedural standardization;
- Liberalization of cross-border movements of money, investments, goods and services (but not labour);
- Guarantees of property rights for global capital;
- Legalization of global organizations and activities.

While some of these regulations have been put in place freely, in some situations countries are under pressure from a variety of sources to implement them (Scholte 2000). Prasad *et al.* (2003) examined the relationship between financial globalization and growth in developing countries. In principle, financial globalization could influence the determinants of economic growth through the augmentation of domestic savings, reduction in the cost of capital, transfer of technology from developed countries and the development of domestic financial sectors. Indirectly, the benefits could be felt through increased production specialization owing to better risk management, along with improvements in macroeconomic policies and institutions induced by the competitive pressures of globalization (Prasad *et al.* 2003). While the average per capita income in more financially open developing economies grows at a more favourable rate than in less financially open economies, whether it is a causal relationship remains a question (Prasad *et al.* 2003). Prasad *et al.* (2003) stress that it is important to understand other factors, such as governance and rule of law, but also acknowledge that some countries with capital account liberalization have experienced output collapses connected to costly banking or currency crises (e.g. the Asian financial crisis of 1997).

Multinational tourism corporations seek to make profits across political boundaries, and it is often the price of production of the holiday, including wages, that is the driving force. Services such as hotel bookings are contracted out, and money or profit is moved from country to country and invested in money markets. Commenting on future trends in the tourism industry, Costa and Buhalis (2006) suggest that rapid liberalization and deregulation of markets will bring fierce competition, which is based not only on price but also on the quality and characteristics of the products supplied. A tourist seeking a vacation in a sunny and warm destination in the Caribbean will often base their decision on price. The idea of increasing competition and globalization in tourism was also commented on by Mowforth and Munt (2003: 12) whereby 'globalisation is about capitalising on the revolutions in telecommunications, finance and transport, all of which have been instrumental in the "globalisation" of tourism'. Those who are able to control or participate in these revolutions benefit while others are left behind (see Box 3.1 for a discussion of globalization and Cancún).

Box 3.1

Globalization and Cancún, Mexico

Cancún, built on a 14-mile-long island on the northern tip of the Yucatán Peninsula, is a large-scale resort complex with approximately 26,000 accommodation units. Financed by the Mexican government and the Inter-American Development Bank, it was built as a regional development project in an area of everglades, swamp and jungle to create new jobs outside traditional urban centres such as Mexico City. The Mexican government granted major regulatory and tax concessions to investors in the early development phase and now Cancún has rapidly become an urbanized resort. Many of the international hotel chains are present, with foreign investment from countries such as the United States, Canada, Italy and Spain. Other chains, such as Pizza Hut, McDonald's, Subway, KFC, TGI Fridays, Outback and Walmart, are present. Cooper (2003) argues that there has been intense competition and that notably Spanish hotel chains have been driving down prices of hotel rooms and cutting costs to squeeze out the competition. They have been offering 'all-inclusive' deals comprising air fare, luxury hotel room with all meals and tips for $50–$60 US a day, which is less than half of the rate on a market peak. The all-inclusive policy has drastically reduced the income of the workers who depend on tips. Other hotels have had to reduce their prices and local suppliers are not being paid. 'Low wages, unstable markets, racism, a high cost of living and poor housing are some of the conditions would-be migrants have found in Cancún' (Hiernaux-Nicolas 1999: 139). The initial infrastructure, built for the surrounding community to house workers, has been well surpassed as the population has grown to approximately 500,000. Migrants who have moved into the area to seek work in the hotels or construction have set up shanty towns. There are also concerns over environmental damage and ground water contamination as well as political corruption (Cooper 2003). The area did sustain major damage by Hurricane Wilma in October 2005 and the reconstruction process is still underway. What was set up as a planned resort has evolved into an urbanized centre that is now exposed to the forces of global competition where price is an all-important factor.

Sources: Hiernaux-Nicolas (1999); Cooper (2003).

As firms move to position themselves in a globalized marketplace, Papatheodorou (2006) notes that a market dualism has emerged in tourism. There is a multitude of small producers (competitive fringe) that coexist with a small number of powerful transnational corporations. The size of these powerful corporations is important in terms of the scope and scale of economies and the imposition of barriers to market entry and exit through asset specificity and irreversibility (Papatheodorou 2006). The expansion and concentration evident in the tourism industry is cause for concern, as it is consistent with anti-competitive practices (Papatheodorou 2006). Several sectors of the tourism industry will be explored briefly below to illustrate trends in the industry.

Airlines and cruise ships

The airlines have undergone significant deregulation with market economies and subsequent competition coming to the forefront of aviation policy making

(Papatheodorou 2006). An increasing number of governments have privatized national airlines to reduce public expenditure and to encourage operating efficiency (Graham 2006). The trend towards deregulation started in 1978 in the US domestic market and similar trends occurred worldwide through bilateral and multilateral agreements (Graham 2006; Papatheodorou 2006). One of the major advancements in deregulation has been the rise in low-cost carriers, which has given new momentum and triggered price wars (Papatheodorou 2006). As with many of the transport sectors there has been a trend towards concentration or horizontal integration with key perceived cost savings in economies of scale, marketing advantages and technological development opportunities (Graham 2006). An example of this is larger airlines being able to capitalize on their networks through strategic alliances with other airlines. Table 3.1 lists the members of three of the largest airline alliances including Star Alliance, Oneworld and Sky Team. The largest, Star Alliance, has 18 member airlines with 2,800 planes serving 842 airports in 152 countries. It carries 425 million passengers a year and employs 360,000 staff (Star Alliance 2006). The Frequent Flyer Programmes that exist within these alliances, inducing passengers to fly with airlines that have global networks rather than using smaller or cheaper airlines, suggests that these partnerships can also create fortress airport hubs where new carrier entry is impossible (Papatheodorou 2006). In a similar vein, globalization is at work in the cruise ship industry (see Box 3.2 and Plate 3.1).

Table 3.1 Members of Star Alliance, Oneworld and Sky Team, 2006

Star Alliance	Oneworld	Sky Team
Air Canada	American Airlines	Aeroflot
Air New Zealand	British Airways	AeroMexico
ANA	Cathay Pacific	Air France
Asiana Airlines	Finnair	KLM Royal Dutch Airlines
Austrian	Iberia	Alitalia
bmi	LAN	Continental Airlines
Lot Polish Airlines	Qantas	CSA Czech Airlines
Lufthansa	Aer Lingus (to leave alliance in 2007)	Delta
Scandinavian Airlines		Korean Air
Singapore Airlines	*Additional Partners in 2007*	Northwest Airlines
South African Airlines	Japan Airlines	
Spanair	Malév	
Swiss	Royal Jordanian	
TAP Portugal		
Thai		
United		
US Airways		
Varig		

Box 3.2

Globalization and the cruise ship industry

In his analysis of the cruise ship industry, Wood (2004) argues that the industry represents the coming of age of neoliberal globalization. The industry is growing and there are predictions in the Caribbean that the cruise ship market share will surpass land-based stay-over tourism. There are no Caribbean-owned ships that cruise the Caribbean sea. Carnival Corporation, which acquired Princes Cruise Lines in 2003, together with Royal Caribbean International, account for 78.7 per cent of passenger capacity in the Caribbean. While both companies had corporate headquarters in south Florida, Carnival is incorporated in Panama and Royal Caribbean is incorporated in Liberia. This has been referred to as flying flags of convenience. Under international maritime law, enforcement of labour, environmental, health and safety laws are done under the flag the ship is flying. Countries that make it known that enforcement and fees will be minimal for such issues as pollution violations are known as 'flags of convenience states' (Wood 2004). Flying flags of convenience and operating at times in international waters offers the industry greater flexibility than land-based operations. One aspect of globalization is minimizing costs and the cruise lines have been criticized for their labour policies. They are able to recruit their labour force from all over the world and pay wages which are a small fraction of prevailing wages in the Caribbean. As the ships have no permanent home, they are free to add to or remove stopovers from their itineraries. Caribbean islands are left competing, trying to attract the ships by offering low port passenger fees. There are debates over the economic impact cruise ships have on destinations and whether cruise ship passengers can be converted to return to the island they have visited as a stay-over guest. As the ships continue to grow in size, ports need to be adapted to accommodate them. Concerns have been raised over the environmental impacts of cruise lines through damage caused by anchors on coral reefs, cases of illegal dumping at sea and the impacts of thousands of cruise tourists disembarking for a few hours in a port of call. Wood notes, however, that the assistance which Caribbean intergovernmental associations have received from the World Bank and the European Union in dealing with cruise ships along with efforts of several NGOs concerning environmental and labour issues are promising developments.

Source: Wood (2004).

Accommodation, tour operators and restaurants

An example of a corporation that has both horizontal and vertical integration across national boundaries is the Accor Group. Based in Europe, it is one of the largest corporations in the tourism industry. It operates close to 4,000 hotels worldwide, ranging from economy to upper-scale service hotels. The company portfolio also includes restaurants, casinos and travel agents, and the firm also provides a range of corporate services (Accor 2006). The hotels in the company's portfolio include Sofitel (upper scale), Novotel, Mecure, and Suitehotel (upperscale and mid-scale), Ibis and Red Roof Inns (economy), Studio 6 (economy extended stay), and Motel 6, ETAP and Formule 1 (budget). Strategies used by hotel chains to extend their reach across the globe include acquisitions, mergers and joint ventures, along with one or more franchises and management contracts and consortia (Go and Pine 1995) (see Plate 3.2). Vertical integration is becoming increasingly common as the same company will control tour operators, retail agencies, airlines

Plate 3.1 ***Nassau, Bahamas: Multiple cruise ships in port.***

Plate 3.2 ***Indonesia, Lombok: Sign indicating where a future Holiday Inn will be built.***

and hotels, thereby securing the inventory and distribution system (Buhalis and Ujma 2006). Tour operators are another sector of the industry that has seen consolidation. In the European market, the four major companies include Thomson (TUI), Airtours (MyTravel), Thomas Cook (JMC) and First Choice, and they are responsible for 80 per cent of package holidays, 66 per cent of flights and over 50 per cent of retail distribution (Reynolds 1999 in Buhalis and Ujma 2006), further illustrating the power of vertical integration. Transat A.T. Inc is an integrated company specializing in organizing, marketing and distributing vacation travel and packages. The company includes tour operators based in Canada and France which have 15 aircraft serving 90 destinations in 25 countries. In the winter, the majority of its charter flights are from Canada to the Caribbean and the United States. The company also sells value added services in destinations as well as distributing travel products through travel agency networks (Transat 2006).

Much attention in globalization has been paid to airlines, hotels, tour operators, and fast food restaurants, such as McDonald's, in terms of expansion and investments abroad. Another example is the Hard Rock Café chain. In a July 2006 edition of an easyJet inflight magazine a single-page advertisement for the Hard Rock Café appeared with the slogan 'You Know Who You Are. You Know Where To Go.' One-third of the page is taken up by the locations of Hard Rock Cafés around the world. Table 3.2 displays the locations of Hard Rock Cafés in developing and transitional countries. The investment decisions taken by the Hard Rock Café clearly illustrate a connection to capital cities and well known tourism destinations.

Table 3.2 Locations of Hard Rock Cafés in developing and transitional countries

Mexico	**Middle East**	**Asia**
Central America	Bahrain	Bali
Acapulco	Beirut	Bangkok
Cabo San Lucas	Cairo	Beijing
Cancún	Dubai	Guam
Cozumel	Hurghada	Hong Kong
Guadalajara	Kuwait	Jakarta
Mexico City	Sharm El Sheik	Kuala Lumpur
Panama		Makati
Puerto Vallarta	**South America**	Mumbai (2006)
Tijunan	Belo Horizonte	Pattaya
	Bogotá	Saipan
Caribbean	Buenos Aires	
Cayman Islands	Caracas	**Europe**
Nassau Bahamas	Rio de Janerio	Bucharest (2006)
Ocho Rios		Moscow
San Juan		Warsaw (2006)
Santo Domingo		

Source: Hard Rock Café (2006).

e-Commerce

The Internet has revolutionized information provision and purchasing patterns in the tourism industry globally, creating not only opportunities but threats to tourism intermediaries (Buhalis and Ujma 2006). As a result of speed and connectivity, boundaries between various channel players are becoming blurred, with some travel intermediaries disappearing while other new ones are emerging. Distribution channels are being revolutionized from linear proprietary electronic distribution systems with a focus on CRSs (computer reservation systems) and GDSs (global distribution systems) to one that is more dynamic and dominated by websites of suppliers, travel agents, web intermediaries, CRSs, GDSs and DMSs (destination management systems) (O'Connor *et al.* 2001). The Internet has allowed consumers to package their own trips by collecting inventory from various suppliers supporting B2C (Business to Consumer) e-Commerce (Buhalis and Ujma 2006). Online travel agencies, such as Expedia.com or Lastminute.com, have emerged offering multiple products from various principals. Suppliers around the world are setting up their own websites avoiding traditional GDSs. An example was provided at the beginning of this chapter of a single tourist on a mass package tour to a developing country to illustrate the link between tourism and globalization. There is no doubt that the mass package tours still dominate the market and the companies that operate them hold a great deal of power. However, if the example is shifted to focus on a traveller on an independent holiday, the diffusion of information communications technologies (ICT) has provided an opportunity for small tourism providers in developing countries to launch dynamic Internet sites attracting the more independent traveller. This development has started to challenge the hitherto unique position of tour operators regarding packaged products (Buhalis and Ujma 2006).

Firms are using a variety of strategies to generate profits in a highly competitive, dynamic and global environment. As Papatheodorou (2006) suggests, there is a dualism evolving in the market with a small number of dominant players. Even on the Internet, mergers have occurred among online travel agents so that the bulk of online travel sales is concentrated within a few companies (Buhalis and Ujma 2006). A question that needs to be examined is to what degree tourism companies in the developing world are able to participate on the Internet and, if they do, whether their websites are picked up through the main search engines. One of the main criticisms with respect to this dualism is that, for developing countries, the control of multinational tour companies is often located in the developed world and profits are often repatriated, leaving little for the destination country.

Political aspects of tourism and globalization

The political dimensions of globalization and tourism are very much interconnected to the economic dimensions. As global corporations wield global economic power, they too wield global political power. This section will focus on interrelated concepts, including the changing nature of the state and the perceived loss of national control in the face of globalizing markets, along with a brief discussion of the political economy of tourism. One of the traditional criticisms of globalization is that as nations open their doors to global markets, state interest is often pushed to the side in favour of the priorities of multinational firms. Countries that have embraced, willingly or otherwise, the liberalization agenda embedded within globalization will have to open their doors to foreign competition. Those countries that have signed up to the World Trade Organisation or have borrowed money from international financial organizations face, to differing degrees, measures that seek to reduce the role of the state. In his book *Confessions of an Economic Hit Man*, Perkins (2004) takes the argument to an even more controversial level, suggesting that developing countries are granted loans which place them in debt to global banking corporations based in the developed world that then require the developing countries to support Western policies on a variety of issues.

The globalization process has impacted on governance. Sovereign states are key to international political regulation; however, governments face a situation where policy options are influenced by other governments or international organizations, especially in the area of economic policies (Lane 2005). Scholte (2000) suggests globalization has facilitated the emergence of post-sovereign governance whereby the state is still important; however, there has been a change in the main attributes. With the rise of supraterritoriality there has been a move towards multilayered governance involving substate, state and suprastate agencies (Scholte 2000). This is evident through organizations such as regional trading blocs and the Bretton Woods institutions mentioned previously in this chapter, and transworld organizations such as various agencies of the United Nations. Some of these organizations have exercised influence over tourism development through funding, such as the United Nations Development Programme (UNDP) and the European Union. Daher (2005) notes the contributions of the World Bank and the Japanese International Cooperation Agency as international donors to various urban regeneration/heritage tourism projects in Jordan. Daher (2005) comments that the funds were channelled through complicated donor agencies' tendering procedures; yet modest outcomes were achieved while foreign debt continued to accumulate. One other change of note with respect to the alterations of the role of the state is privatized governance whereby there is regulatory growth occurring outside the public sector (Scholte 2000). This involves, for example, a range of business associations, NGOs, think-tanks, foundations and even criminal syndicates (Scholte 2000).

The changing nature of governance and the rise of governing structures, which stretch across political boundaries, raise issues of power and control. We will now examine the political economy of tourism. Early discussions of the political economy approach to tourism focused on the idea that tourism has evolved in a manner that closely resembles the historical patterns of colonialism and economic dependency: 'Metropolitan companies, institutions and governments in the post colonial period have maintained special trading relationships with certain élite counterparts in Third World countries' (Lea 1988: 12). These groups from the developed world, through their close association with local political and commercial classes, are able to encourage decisions over economic policy, labour legislation and commercial practices that are consistent with their interests (Britton

1982). The overall tone of political economy tends to be negative about the impacts of tourism, as it is a way for wealthy metropolitan nations to develop at the expense of less fortunate nations (Lea 1988). Power and control issues raised by the political economy approach to tourism may also be observed within dependency theory as outlined in Chapter 1. National economic autonomy in the developing world becomes subordinate to the interests of foreign pressure groups (including multinational tourism companies) and privileged local classes rather than from the development priorities arising from within a broader political consensus in the developing country (Britton 1982, 1991). Britton (1982) developed an enclave model of Third World tourism comprising a three-tiered hierarchy. At the top are the metropolitan market countries where the headquarters of the major tourism companies such as transport, tour companies and hotels are located, and they dominate the lower levels in the hierarchy. In the centre of the model are the branch offices and associated commercial interests of the metropolitan companies located in the destination that operate in conjunction with local tourism counterparts. At the base of the pyramid model are the small-scale tourism enterprises which are marginal but dependent upon all of the organizations located higher up in the model. Foreign companies then come to dominate the system and capital accumulation travels up the hierarchy so that the multinationals benefit the most.

In reviewing the literature, Cleverdon and Kalisch (2000) suggest that far from bringing economic prosperity to developing countries, tourism has great potential to reinforce economic dependency and social inequality. While some fortunate sections of society, such as ruling élites, landowners, government officials or private businesses may benefit, the poor, landless and rural societies are getting poorer not only materially but also in terms of their culture and resources. Examples noted by Cleverdon and Kalisch (2000) include eviction and displacement for resort construction, rising land, fuel and food prices, and commoditization of culture. Lea (1988) outlines four main reasons for the successful invasion of developing countries: (1) they seldom invest large amounts of their own capital in developing countries but rather seek funding from private and government sources in the destination; (2) the infrastructure required to support new resorts (e.g. roads, power lines) are also funded locally or through foreign loans; (3) a viable visitor flow is maintained through global marketing campaigns; and (4) transnational corporations share profits through charging management fees, limited direct investment, and various franchise, licensing and service agreements. Their power comes from their ability to withdraw from these agreements, leaving the developing country in a very vulnerable position. Drawn by the attractiveness of the tourism industry and its perceived economic windfall, governments actively pursue multinationals to set up operations in their countries. This has led to countries being forced to enter a bidding war, offering a range of things such as forgivable loans, reduced taxes or complete tax holidays, along with the failure to enforce environmental laws in the hopes of attracting the multinational (Reid 2003). In an interview with a hotel manager of a beach resort in Tunisia in 1996 by one of the authors of this book, the manager indicated that the demands of the tour operator with whom they had been working for the past several years for extra services had been increasing, yet the payment made to the hotel had not increased.

While the traditional model of the political economy of tourism tends to focus on the controlling actions of multinational companies in developed countries, Bianchi (2002) argues that through economic globalization and market liberalization there is an increasingly complex and differentiated geography of tourism production. This dynamic change in geographic production challenges the straightforward north–south power and control balance presented in the neo-colonial/dependency model of international tourism. The challenge to the state-centric approach is the increasing dominance of transnational tourism corporations, and the growing structural power of market forces at global and regional levels (Bianchi 2002). The international political economy (IPE) approach argues that orthodox international relations work is deficient, as it focuses exclusively on the relations of the government (Preston 1996). The IPE approach draws on development economics, historical sociology and economic history, and one of the key ideas is that the basic needs of any polity are related to wealth, security, freedom and justice (Preston 1996). In summary, Preston (1996: 291) suggests: 'most broadly, the IPE approach offers the model of a world system comprising a variety of power structures within which agent groups, primarily states, move and where the specific exchanges of agent groups and global structures generate the familiar pattern of extant polities.' Regionally based economies such as the Singapore, Indonesia and Malaysia Growth Triangle (Telfer 2002b) have become more autonomous actors competing for mobile tourism capital (Bianchi 2002).

While there is a great deal written about the dominant forces of globalization, it has also been argued that it is too simplistic to accept globalization as an 'all-encompassing unilateral and hegemonic force' (Teo 2002: 459). Hazbun (2004) argues that many of the studies of globalization focus on the loss of the power of the state, resulting in deterritorialization. However, Hazbun (2004) argues (using examples from the Middle East) that it can also lead to reterritorialization where the state can exercise power and control over territorial assets. Using the example of Tunisia, the country has diversified from beach tourism to promoting its urban and Islamic heritage as well as creating a new market by focusing on its less developed southern desert region. The desert product was an attempt to develop a territorially specific place image that had previously identified with a low-cost Mediterranean beach destination. The strategy included marketing a more exotic image of Tunisia, with its urban markets and desert landscapes, where movies such as *Star Wars* and *The English Patient* were filmed. While acknowledging the difficulties, Hazbub (2004: 331) suggests that by 'asserting control over tourism spaces and processes that convert places, cultures and experiences into territorial defined tourism commodities', actors such as states, local communities or transnational corporations are able to drive the reterritorialization and thereby control the conditions that draw tourists and capital investment.

Global and local forces do interact in the destination. There are a number of authors who suggest that it is important to recognize the global–local nexus (e.g. Robins 1997; Teo 2002). As Robins (1997) argues, the global economy cannot simply override existing social and historical realities. Global entrepreneurs need to negotiate and come to terms with local contexts, conditions and constraints. As the globalization of the tourism industry continues, more parties are becoming involved. Each group, whether it be tourists, hotels, travel agents, private enterprises, local governments, NGOs or individuals, has to protect its interests (Teo 2002). Globalization has established new and complex relations between global and local spaces (Robins 1997). The key question is, what is the strength of the local institutions with respect to the forces of globalization? This will be explored in the next section.

Cultural aspects of tourism and globalization

The stereotype of cultural globalization has Western lifestyles and forms of consumption spreading across the globe, resulting in a convergence of culture that is defined by capitalism (Allen 1995). The process has also been referred to as cultural imperialism and can lead to assimilation. The so-called 'global products', such as Coca-Cola, Levi jeans, McDonald's hamburgers or Benetton clothes, are reflections of the idea that as people gain access to global information they become globally aware (Robins 1997), and tourism is one source of global information. The global media send idealized images of destinations to potential tourists in the developed world, while those in developing countries receive images of Western consumption patterns. Advances in technology and communications have taken people to more remote areas on adventure or ecotourism holidays. Travel writer D'Amico (2005) refers to the 'Coke Line', which is the line one crosses to find a community not yet touched by outside forces. Specifically, he is referring to indigenous communities that have not yet been exposed to Coca-Cola. This line is rapidly disappearing. The further people travel the fewer and fewer places there are left that have not been exposed to tourists. The Chinese government has spent US $4.2 billion constructing a train route into Lhasa, Tibet. The Sky Train travels at such a high altitude (3,700 metres) that the Canadian-made rail cars are pressurized like an airline. Proponents of the plan argue that it will help the economy of Tibet, including doubling its annual tourism income to US $725 million by 2010. Critics, however, raise concerns over environmental damage and the impact this will have on the culture of Tibet (Olesen 2006). How communities and individuals react to outside forces may be an indication as to the strength of the host culture. Is there a resilient local culture or will it be overrun, not only by the culture of the tourists but also by the rapidly evolving global information networks? Part of the tourism development dilemma associated with culture is that if a culture is frozen in time to be put on display for tourists, can development still occur? Cultural brokers will seek to make a profit while others will try to avoid the tourists. The remainder of this section will focus on different aspects of culture, tourism and globalization.

Cross-cultural contact

Tourism brings together cultures from around the world and a variety of studies have been conducted on the nature of guest–host relationships (Smith (1977) is an important early work), and the social and cultural impact of tourism (Wall and Mathieson 2006; see also Chapter 7). Reisinger and Turner (2003) have explored cross-cultural behaviour in detail. The recent launch of the *Journal of Tourism and Cultural Change* is also a reflection of the interest in the area. Culture is a highly contested concept with many differing definitions. Peterson (1979) suggests that culture consists of four kinds of symbols that include values (choice statements ranking behaviour or goals), norms (specifications of values relating to behaviour in interaction), beliefs (existential statements about how the world operates, which often serve to justify values and norms) and expressive symbols (any and all aspects of material culture). To what extent does tourism and the processes of globalization change any of these four 'symbols' of culture? A number of concepts have been explored in the tourism literature on cultures coming together, including assimilation (where one culture loses out and is brought into another culture), acculturation (where cultures come together and share attributes) and cultural drift (where hosts temporarily adjust their behaviour while in contact with tourists) (Wall and Mathieson 2006). Since tourism is inevitably linked to globalization, does the host culture take on attributes of the tourists' culture, or is there a broader emerging global culture that tourism is helping to spread? Caution, however, must be used when considering a convergence towards a global culture. Robins (1997) suggests that there are three main aspects to consider in terms of culture and identity with respect to globalization. The first is homogenization whereby cultural globalization brings about the convergence of cultures. The second is that there are cosmopolitan developments with cultural encounters across frontiers resulting in cultural fusion and hybridity. The third is a reinforcement of cultural diversity, often against the perceived threats to homogenization or cosmopolitanization. There are those who react to global changes by returning to their roots and reclaiming their traditions, thereby reinforcing the diversity of cultures. The reclamation of tradition may stimulate cultural tourism, as will be discussed below. The impact of globalization on cultures is complex, pulling cultures in different, contradictory and conflicting ways (Robins 1997).

Conflicts can result from clashes in culture, religion and family values as shifts occur in the structure of the labour force. In the context of tourism, much has been written on the demonstration effect associated with international tourism. Tourists model certain types of behaviour through action and dress, which are copied by those living in the destination. A fisherman-turned-supplier for the Sheraton Senggigi Beach Resort in Lombok, Indonesia changed his manner of dress from the traditional Indonesia sarong to Western-style clothing. At the same hotel, a local female hotel employee was prevented from returning to work by her father after only a short time, as he could not get used to his daughter mixing with foreign tourists and making more money than him. Workers in the tourist industry also typically gravitate towards the language used by tourists.

Tourism results in immigrants moving into the destination to take up jobs in the tourism industry and bringing along their own culture. There is concern in the Galapagos Islands that migrants from Equador's mainland, hoping to get a job in tourism, are putting a strain on the island and its residents already dealing with an influx of tourists (Schemo 1995). These migrants may also bring their own customs and traditions which interact with the local culture (Bookman 2006). The conditions of employment in developing countries have raised concerns. Tourism

Concern, a British NGO, is currently running a campaign against employment conditions at hotels in developing countries where many employees do not have contracts and are trapped in poverty through low wages. In some developing countries, sex tourism is part of the tourism product and this is emphasized through the global media where residents in the destination are presented as the desirable exotic 'other' (Momsen 2004). Mitchell (2000: 88) explores the political economy of culture and suggests that 'geography – spatiality, the production of spaces and places, and the global doings of the economy – is front and centre in the processes of cultural production, and hence is systems of social reproduction'. This raises the issue of how cultures are portrayed around the world.

As globalization has facilitated travel and interactions between hosts and guests, there is also concern over the rapid spread of infectious diseases. In 2002 SARS (Severe Acute Respiratory Syndrome) originating in Guangzhou, China quickly spread to a number of countries. As Kimball (2006: 45) suggests, 'the unknown virus was going global, travelling by airline'. Tourists who participate in sex tourism are at risk of contracting a variety of sexually transmitted diseases. Kimball (2006) argues that with increased travel and trade there is clearly a risk with globalization that is creating a new ecology of infectious diseases.

Interactions between people of different cultures can also have a positive outcome whereby knowledge and understanding are generated among the groups. New ideas on topics such as empowerment, human rights and peace may be shared, thereby contributing to broader notions of development. Macleod (2004) raises the issue of identity in the context of tourism and cultural change. He suggests that it is valuable to consider people as having a complex set of identities that form the overall picture. Where tourism occurs, people may experience change in identity on a personal and public level as individuals and groups. Fishermen become shopkeepers or tour guides, young women seek financial independence or pressure groups revitalize identities. Macleod (2004) goes on to suggest that in the context of globalization, people may encounter a broader range of role models, networks of associates and opportunities, as well as undergoing a new self-perception. People or groups may also experiment with traditional roles, reinterpret events, reinvent themselves or gain confidence in their lifestyles and values (Macleod 2004: 216).

Cultural tourism

Destinations present themselves through many different cultural factors such as 'entertainment, food, drink, work, dress, architecture, handicrafts, media, history, language, religion, education, tradition, humour, art, music, dance, hospitality and all the other characteristics of a nation's way of life' (Reisinger and Turner 2003). People travel to either see or immerse themselves in 'authentic' local culture and heritage tourism is receiving greater attention in the literature (Daher 2005; see also Chapter 4). Some argue that culture and tradition can be saved through dance, art, festivals and souvenirs, while others claim that this commoditization of culture for the tourists destroys the original culture. Culture sells, and it is packaged or commodified for the tourists, raising debates about what is authentic. Crafts are modified so that they are attractive and portable for tourists. This shift, however, may very well generate a great deal of income and hence lead to development. Azarya (2004) examined the challenges and contradictory consequences of indigenous groups being part of the tourism industry with reference to the Massai pastoral groups in Kenya and Tanzania. The positive outcomes were associated with new sources of income through working in the industry, as well as exhibiting themselves as tourist attractions, being photographed, selling souvenirs, opening their villages and putting on dances. These benefits come at the cost of 'freezing' themselves at the margins of society as it is their marginality which is the essence of the tourism product.

Tourist cultural enclaves

While some tourists seek out the local culture when they travel, others want to avoid it and choose to remain inside the confines of their hotel. A tourist cultural enclave may emerge for those who choose to remain within the resort area and there is therefore very little contact with the host community. A lifestyle is transplanted from the developed world to the developing world and this provides few opportunities for locals to make money from the tourists if they remain behind the walls of the resort. From a solely economic perspective, it may be in the tourism resort's best interests to keep their tourists inside the resort so that they will spend money in the restaurants and on souvenirs rather than face the competition from outside the doors of the hotel. From a planning perspective, the development of enclaves may be a conscious choice in order to limit cultural contamination through tourist–local encounters.

Business culture

One other area to be explored in terms of culture is the influence of globalization on business culture. Previously in this chapter, examples of various multinationals have been given, illustrating the global reach of these firms. These firms also have their own operating procedures which, in most cases, contain elements of standardization across the multinational chain, not only for efficiency reasons but also for the tourists' peace of mind so that they can expect the same level of service from hotel to hotel operating under the same brand. The question is whether the multinationals are able to adapt their standardized operating policies and procedures to fit within the local destination. For example, the restaurant chain McDonald's adapts its menus, and hotels will often incorporate local architectural designs into their construction. While this is relatively easy, what may be more of a challenge is to incorporate Western-style management systems into non-Western settings.

Globalization has not only opened up the cultures of the world to tourists; it has also opened up cultures of the world to residents of the destinations. Globalization has the potential to magnify cultural impacts, as so many groups are moving around the world and people are rapidly exposed to mass media. Culture is however dynamic and, as Wall and Mathieson (2006) suggest, cultural change is influenced by both internal and external factors, and cultures would change in the absence of tourism. Culture is not only an attribute of a destination it is also a product, and therefore culture in its various forms has the potential to have a significant impact on tourism planning and development, management and marketing (Reisinger and Turner 2003). These interactions between tourism and culture are explored further in Chapters 5 and 7.

Conclusion

Globalization is a highly complex and multifaceted process; however, stereotypes associated with globalization are often oversimplifications of a very dynamic process. Globalization is not a fair process, and there are winners and losers. In reflecting on globalization, McMichael (2004: 152) states: 'it is tempting to think of globalisation as inevitable, or as destiny, until we notice that it only includes one-fifth of the world's population as beneficiaries.' The dilemma for developing countries is whether or not they decide to open their doors to let the multinational tourism companies into their country. However, do they really have a choice? The international companies bring a recognizable brand and image associated with them along with the necessary business infrastructure to attract international tourists. It does come at a cost, however. The proponents of the global capitalist system have argued that the growing world economy in which tourism is a major player (although not always readily acknowledged) is vital to global development. However, just as critics have signalled the end of development as a concept, so too are they raising the spectre of the collapse of globalism (Saul 2005). While acknowledging the growth in global trade as a success story, the promises of globalization are not materializing, as there is ongoing Third World debt and poverty (Saul 2005). Tourism is 'quintessentially linked to globalisation and the phenomenon of time–space compression. In the sociocultural realm, tourism is emblematic of globalisation, hyperreality, fantasy and post modernity' (Potter *et al.* 1999: 95). If tourism is to be used as an agent of development, then there are calls for the implications of globalization for destinations to be studied less as negative versus positive but more as a process that is dynamic, contingent and contested (Teo 2002). The following chapter considers the tourism development process which is highly influenced by globalization.

Discussion questions

1. **What power do developing nations have in the face of multinational tourism corporations?**
2. **What is the role of information technology in supporting the controlling nature of multinational tourism corporations?**
3. **Can you identify ways in which multinational corporations can improve their linkages with destination communities?**
4. **Critically evaluate the neocolonial dependency model in relation to tourism development.**
5. **What is the impact of international tourism on local culture?**

Further reading

Hall, C. M. (2005) *Tourism: Rethinking the Social Science of Mobility*. London: Pearson Prentice Hall. [This book provides an excellent account of key concepts related to mobility. Tourism is examined through such concepts including globalization, localization, identity, security, and global environmental change.]

Stubbs, R. and Underhill, G. (eds) (2006) *Political Economy and the Changing Global Order* (3rd edn), Don Mills, Ontario: Oxford University Press. [This book is structured into the following four sections: (1) changing global order, (2) global issues, (3) regional dynamics, and (4) responses to globalization. There is a good analysis of the power and control issues in globalization.]

Websites

The website of the InterContinental Hotel Group illustrates the nature of a global tourism corporation. There is a wide variety of information on the various brands within the company and investor information: http://www.ihgplc.com/.

The Star Alliance is an alliance of airline companies with global reach. Background information on the formation of the alliance and information on the partners are available on this website: http://www.staralliance.com/en/travellers/index.html.

The website of the United Nations Environment Programme (UNEP) activities in tourism. The site outlines the organization's sustainable development initiatives and illustrates how a global-level organization can influence tourism development at state level: http://www.unep.fr/pc/tourism/home.htm.

4 The tourism planning and development process

Learning objectives

When you have finished reading this chapter, you should be able to:

- **Understand the tourism planning and development process;**
- **Be aware of government policy on tourism developments;**
- **Identify the changes occurring in tourism planning;**
- **Debate the pros and cons of various forms of tourism development.**

Tourism development is a complex process involving the coming together of domestic and international development agents and key stakeholder groups with state policy, planning and regulations. The resulting tourism form not only has impacts in the host destination but there are also potential broader developmental outcomes benefiting that destination. Cruise ship docks, beach resorts, urban heritage centres, national parks, ecotourism resorts, casinos and village tourism represent a few of the very different and diverse tourism products created to attract tourists to developing countries. Decisions need to be made as to what form or forms of tourism are best suited to a destination for the long term in order to meet national developmental goals. For example, does concentrated large-scale mass tourism development, with high levels of multinational involvement such as a beach resort, do more in achieving developmental goals, or does small-scale dispersed community-based tourism generate more benefits? In a highly competitive global market, most developing countries are choosing policies to diversify their products. Cuba, for example, in addition to its beach resorts, is rapidly expanding its ecotourism sector. Previous debates on form and function tended to be polarized around the idea that one form of tourism is more sustainable than another, and more recently it has been recognized that all forms of tourism need to be made more sustainable. Each type of development will bring different tourists, with varying levels of disposable income and expectations, along with different opportunities for locals to participate in the tourism economy. State policy and planning on tourism will often dictate the nature of development in a destination and, given the complexities of the development process, what works in one destination may not work as well in another. In order to address the questions of form and function of tourism, one needs first to examine a series of politically oriented questions that are central to the tourism and development debate.

- What is the desired outcome of the development?
- What are the tourism policy and planning regulations in the destination?
- What are the institutional arrangements and the political realities in the destination?
- What are the values of the key actors and institutions involved in the development process?
- Who is in control of the decision-making process?
- What project is selected, how is it financed and who operates it?
- Who benefits from the development?
- Can tourism development contribute to national development goals?

In developing countries, tourism planning and development often occurs through a top-down planning approach. Liu and Wall (2006) argue that decision making with respect to tourism developments is predominately based on the interventions of government agencies and large tourism firms, resulting in the dominance of external, often foreign capital and the marginalization of local people. Within sustainability there are calls for increased local involvement in the planning process; however, it is important to consider to what degree this is possible and how it can be facilitated. The focus of this chapter will be on the overall tourism planning and development process, along with an examination of selected strategies of tourism development as potential tools to help meet the developmental goals of a destination.

Tourism and development process

In the tourism and development debate, one of the first questions to be considered is: What is the desired outcome of any tourism expansion or refurbishment? Gunn and Var (2002) identify the goals for better tourism development as enhanced visitor attractions, improved economy and business success, sustainable resource use, and community and area integration. When the question is asked, why get into tourism?, economic arguments are often presented first, with host governments trying to establish a strategy that has the correct mix of tourism developments to generate the maximum amount of foreign exchange, the argument being that once economic benefits start to accrue then other spin-off benefits will arise for the host society in terms of other developmental goals. It is difficult to say to what extent tourism, pursued from an economic perspective, will lead to broader development goals such as greater self-reliance, endogenous growth, fulfilment of basic needs, environmental sustainability or other goals related to the UN Millennium Goals (see Box 1.1). The selection of any particular tourism development strategy or combination of strategies is a very complex process. It needs to be placed within the context of the politics of place and an understanding of the institutional arrangements in the destination as well as an understanding of the values and power of those with the investment funds. There has to be the recognition that, in some cases, immediate profitability from a new cruise ship dock or resort may take precedence over environmental protection or other broader development goals, and governments are often forced to make a trade-off between competing concerns. Other destinations may adopt a tourism policy guided by sustainable development and, as part of that policy, reduced numbers of visitors may well be a target. These potential trade-offs illustrate the difficulty in implementing sustainable development, and echo comments by Mitchell (1997) that change, complexity, uncertainty and conflict are central in resource and environmental management and, it is argued here, they are also central in planning tourism development.

Figure 4.1 illustrates the complex nature of the tourism and development process as it relates to form and function. A brief explanation is provided here in advance of more detailed analysis in the remainder of the chapter. The first set of elements to be considered are the values, ideologies, goals, priorities, strategies and resources of tourism development agents. Figure 4.1 contains a list of some of the main tourism development agents, which includes various government, private industry and not-for-profit organizations. Some of those with influence are in the destination, while other organizations operate from abroad, such as multinational corporations, international funding agencies or regional trading blocks, as illustrated in Chapter 3. The interaction of these various groups occurs through a policy, planning and politics filter. This is referred to as a filter here, as any development project must work its way through the layers of political and bureaucratic structures in the destination. The political scope of a tourism development can potentially extend beyond the local political situation, incorporating multiple dimensions and interest groups across local, regional, national and even international scales that may be for or against future development. Policies are established, plans are written, and regulations are put in place by governments to either promote or restrict development. It is in this environment that governments may try to offer various incentives to attract tourism development. Corruption may also subvert regulations. Proposals for tourism development are put forward through the planning and regulatory processes in the destination and, if the project is externally driven, local meets global. Key factors here include the type of political structure that exists in the developing country, institutional arrangements, and to what extent local participation is permitted in the planning process.

What is ultimately physically constructed will be greatly influenced by the tourism policy of the state, the values of those in charge, and the resources they have to work with. The design of a building may relate to whether a destination has been branded or strongly associated with an established type of tourism (e.g. ecotourism in Costa Rica, safari tourism in Kenya or beach tourism in Fiji). As illustrated in the centre of Figure 4.1, there are debates in the literature over the scale that tourism development should take and the merits of enclave tourism

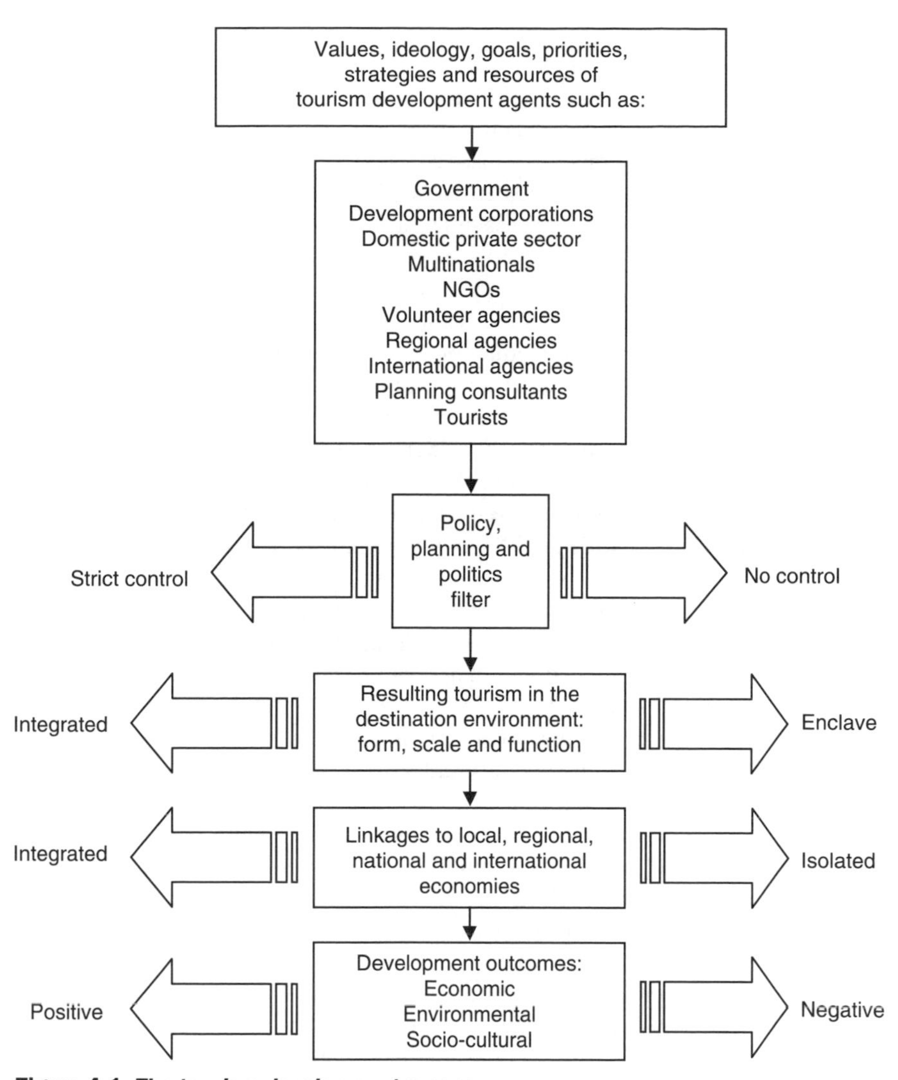

Figure 4.1 ***The tourism development process.***

versus integrated tourism, which typically has more opportunity for links with local communities. Later in this chapter the pros and cons of a number of different tourism developments will be explored in the context of contributing to overriding development goals. One of the important considerations in the overall development process is the strength of linkages to the local economy, both in terms of the formal and informal sectors. Those linkages may be at local, regional or national level; however, as the linkages extend beyond national borders, leakages start to increase. The final section of Figure 4.1 illustrates the overall development outcomes in terms of the economy, environment and society.

Values and power

The first box in Figure 4.1 lists the values, ideologies, goals, priorities, strategies and resources of the various tourism development agents. It is important to consider the interrelationship of these concepts for the various individuals and groups involved in tourism development, as well as their level of power and control in the overall development outcomes in the destination. Goldsworthy (1988) argues that all development theories, policies, plans and strategies either consciously or unconsciously express a preferred notion of what development is, and it is these preferences that reflect values. Values have been regarded as an 'enduring belief that a specific mode of conduct or end state of existence is personally or socially preferable to an opposite or converse mode of conduct or end state of existence' (Rokeach 1973 in Gold 1980: 24). Peterson (1979) suggests that values, then, are choice statements ranking behaviour or goals. Differing development agendas will reflect different goals, and these goals will reflect social, economic, political, cultural, ethical, moral and even religious influences (Potter 2002). As the nature of development has widened its scope beyond economics, the literature on development ideologies has changed to also emphasize political, social, ethnic, cultural, ecological and other dimensions of the broader processes of development and change (Potter 2002). A great deal of development thinking also remains politically uninformed, and more attention needs to be given to the ideological underpinnings of development theory (Goldsworthy 1988). A few examples of broad categories of ideological underpinnings of development theory include positions such as conservative, liberal and radical (Goldsworthy 1988). Peet and Hartwick (1999: 3) suggest that 'developmentalism is a battle ground where contention rages between bureaucratic economists, Marxist revolutionaries, environmental activists, feminist critics, postmodern sceptics, radical democrats and others'. Development has also been criticized for its role in Western imperialism (see Said 1978).

In examining tourism policies in developing countries Jenkins (1980) notes the dichotomy in aims between private investment and government investment. Private investment criteria centre on profitability with other considerations taking secondary place. Government investment, although concerned with profitability, must also consider other non-economic implications such as social and cultural impact and land-use policies. Depending on their priorities, governments may take a 'tourism first' approach, focusing efforts on the industry itself or a 'development first' approach where planning is framed by broader national development needs (Burns 1999a). As mentioned in Chapter 2, with the shift to sustainability there has been increased concern about issues such as corporate social responsibility. Robbins (2001) refers to firms with a traditional corporate culture focused on generating wealth, and firms with a socio-environmental culture where the focus is not only on wealth but also on being socially and environmentally active.

Tourism development agents have values and ideologies that need to be considered within the context of power. There is a wide diversity of approaches to the study, meaning and definition of power (see Coles and Church 2007), and several authors have stressed the importance of understanding power relations in tourism (Hall 1994, 2007a; Mowforth and Munt 1998, 2003; Bianchi 2002; Coles and

Church 2007). Questions such as how much power tourism development agents hold, where they get their power base from, and to what extent they are able to use their power to generate their preferred outcome are important considerations in the study of tourism planning and development. Todaro (1997) argues that most developing countries are ruled directly or indirectly by small and powerful élites to a greater extent than developed nations.

Power comes from a variety of sources and is exercised in different fashions in the tourism development process. In examining social power, Mann (1986) suggests that a general account of societies, their structure and history can best be given by the interrelations of four sources of social power: ideological, economic, military and political. In the context of tourism, governments may have power based on laws and regulations, while industry may have power through wealth, information or technology, and citizen groups may have power through local participation. The institutional arrangements will influence, in part, how power can be used; however, corruption may get around laws and regulations. Duffy (2000), for example, found that in Belize there has been an expansion of organized crime, and the enforcement of environmental regulations, on which ecotourism is based, has become problematic. The government regulations on the environment are 'rendered ineffective when a junior arm of the state is opposed by more powerful interest groups that lie within and outside the state apparatus' (Duffy 2000: 562).

The political economy of tourism was examined in the context of the globalization of the tourism industry in Chapter 3. In examining the changing global order, Underhill (2006) suggests that the most recent expression of the complex relationship between power and wealth (or between the political and economic domains) in international politics are the links between political authority, on one hand, and the systems of production and distribution of wealth, referred to as the market, on the other. As for tourism, to what extent do governments set policies to open up their borders to multinational corporations? Bianchi (2002) suggests that conceiving market behaviour in isolation from the ideologies and values of the various actors and groups involved, as reflected in the free market notion of *comparative advantage*, underplays the political nature of markets whereby the state has historically conditioned the activities of the economic classes, and ignores the uneven consequences of unlimited market competition. Bianchi (2002: 268) summarizes the political economy of tourism as 'the examination of the systematic sources of power which both reflect and constitute the competition for resources and the manipulation of scarcity, in the context of converting people, places, and histories into objects of tourism'. As pointed out in Chapter 1, in the discussion of *Beyond the Impasse: The Search for a New Paradigm?*, Hettne (2002) suggests that the international political economy is one of the building blocks in moving discussions forward about development theory. In establishing the background to international political economy, Underhill (2006) reviews the contributions from three main approaches in international relations, including the realist tradition (the state as unified autonomous actor), liberal approaches (interaction of individuals in the economic sphere) and radical approaches (e.g. Marxist), as well as noting the literature on feminism, environmentalism and postmodernism. The exercise of power in the international political economy occurs in a setting characterized by complex interdependence among states, their societies and economic structures both at a domestic and an international level (Underhill 2006). This takes place through an integrated system of governance 'operating simultaneously through the mechanisms of the market and multiple sovereignties of a system of competitive states' (Underhill 2006: 19). As Milner (1991) in Underhill (2006) suggests, examining the international system as a web of interdependencies necessitates a focus on linkages among actors. The chapter now turns to examining various actors in the tourism development process.

Actors in the development process

A wide variety of actors representing the public, private and not-for-profit sectors come together in the tourism development process, as illustrated in Figure 4.1. In this section a variety of examples will be used to illustrate the different roles. As outlined above, each of these actors has a set of values, ideologies, goals, priorities, strategies and resources which they bring to the tourism development process. At times these various groups will have similar end goals, while at other times they may be in conflict with one another. A private developer may seek maximum profitability through the construction of a high-rise resort, while the state may choose to regulate building heights. All actors contribute in some fashion to the supply of tourism products, which Page and Connell (2006) summarize as national and regional tourism organizations, destination organizations, local authorities, attractions, transport, accommodation, the retail sector (shops), travel agents, tour operators, and the hospitality industry (restaurants and cafés).

State involvement

One of the most important actors in the tourism development process is the state. Through various government ministries or state-sponsored tourism development corporations, governments establish the framework, policies, plans and regulations to both attract and control tourism development. As pointed out in Chapter 3, with the rise of globalization there has been a change in the power of the state. In some regards the state has lost a degree of power in the face of multinational corporations; however, in another sense there is a more complex multilayered governance involving substate, state and supra-state agencies (Scholte 2000). In many developing countries the state plays an important role in facilitating tourism and there is a top-down approach to planning (Liu and Wall 2006). The government shapes the economic framework for the tourism industry, provides infrastructure and education requirements for tourism, establishes the regulatory environment in which businesses operate and takes a role in promotion and marketing (Hall 2005). Hall (1994) suggests that there are seven roles played by government in tourism. These are coordination, planning, legislation and regulation, entrepreneurship, providing stimulation, social tourism, and interest protection. In Mexico, the Secretariat of State for Tourism formulates and carries out tourism policy and supervises the operations of Fonatur, the national fund for the development of tourism and the Mexican Tourism Board. Fonatur arranges financing for tourism enterprises along with real estate transactions, while the Tourism Board looks after promotion (See Box 4.1). A key element in the success of tourism development is the policies and regulations put in place to attract both domestic and foreign tourism investment to a country or region. In Mexico, the target of US$9 billion for private investment (domestic and foreign) in the tourism sector for the period 2001 to 2006 was passed in June 2005. Forty-eight per cent of

Box 4.1

Fonatur and tourism development in Mexico

Fonatur is Mexico's National Tourism Development Trust created in 1974 to develop planned integrated resorts and to not only promote regional development but also create projects that will have a national impact. The organization devises master plans for development areas, arranges for financing of tourism enterprises, conducts real estate transactions and looks after infrastructure projects related to tourism development. Fonatur has five integrated resort developments in Mexico including Cancún, Los Cabos, Ixtapa, Loreto and Huatulco Bays. Together these five resorts include over 245 hotels and more than 36,800 hotel rooms. These destinations receive 54 per cent of Mexico's foreign revenue from tourism and 40 per cent of all foreign visitors. They generate US $2.76 billion annually along with $300 million in taxes from sales and lodging taxes. The States of Quintana Roo and Baja California are home to the tourism developments of Cancún and Los Cabos respectively and these two states ranked fourth and eighth in per capita GDP respectively. Fonatur actively seeks out investment for tourism development and provides a favourable environment for foreign investors. The regulatory framework for investment in Mexico supports foreign ownership in most economic fields and activities including real estate, allowing 100 per cent participation in shared capital. In addition to offering legal guarantees and security to foreign investors, the system in place allows unrestricted repatriation of profits, bonuses, dividends and interest payments. Fonatur holds an annual conference to attract developers and investors in tourism. The organization is currently offering a new investment proposal called a 5 × 5 plan. The plan includes (1) beachfront property in all five of Fonatur's resort areas, (2) financing for construction at competitive rates, (3) processing of all licensing and permits needed for the project, (4) fixed price construction estimates, and (5) guarantees in infrastructure and land ownership. In addition to redevelopment and expansion plans for the existing major resorts, Fonatur has a number of other current projects underway to diversify and develop new tourism products, including two mega-projects. The first is the construction of a premium integrated resort in the western coastal state of Nayarit designed to relieve pressure from Acapulco. It is the first time in 20 years that Fonatur has been involved in the development of a new city, which is expected to have 14,500 rooms. The second large-scale project is the three-pronged Sea of Cortez project which is focused on enhancing marina development along the Baja Peninsula. The mission statement for Fonatur, in part, claims that it will be the institution responsible for the planning and development of sustainable tourism projects of national impact. Part of the vision includes a statement that development should occur with a social awareness that favours regional development and creates permanent jobs that are appropriately remunerated. The challenge Fonatur faces is not only in the actual development of these large-scale projects but to ensure that the benefits circulate to as many as possible. As Torres and Momsen (2004) discovered, in the case of Cancún there are numerous challenges to overcome in strengthening backward economic linkages between these large resorts and other sectors of the local economy, such as agriculture.

Sources: Fonatur (2006); Visit Mexico Press (2006a, b); Torres and Momsen (2004).

private investment went into Mexican beach destinations and the United States is the largest foreign investor in Mexico's tourism infrastructure (Visit Mexico Press 2006a). At the other end of the spectrum, the state can provide access to micro-credit schemes to help facilitate pro-poor tourism (see Chapter 5). At a regional or state level, the West Bengal government in India is appointing a consultant to attract Indian and international investors to the state's tourism sector and to formulate future tourism policy. The areas for expansion identified include tea tourism, river tourism, Raj-nostalgia tourism and ecotourism (India eNews 2006). The state also plays a role in destination marketing and creating an image that is attractive to tourists. Howie (2003), however, notes the tensions between the traditional planning approaches, guided through concepts such as carrying capacity, conforming land uses, acceptable functional uses and sustainable development, with the largely economic-driven market perspective of local chambers of commerce, major investors and influential bodies representing the private sector business. The state has to become increasingly entrepreneurial in a very competitive market. There has been a growth in the state entering partnerships with industry or other states in marketing or providing a facilitator role for the formation of partnerships for smaller companies. Marketing is explored further in Chapter 6.

Private sector involvement

Private sector development ranges from small-scale entrepreneurs in the informal sector, such as local tour guides, to small and medium-sized enterprises (SMEs), up to domestic hotel chains and multinational corporations, such as tour operators or hotel companies. It is important to note that while the larger corporations often attract the most attention, the small tourism business and informal sector are crucial in terms of employment and generating opportunities for locals to participate in the tourism industry. These organizations will be explored in more detail in the next chapter. Criticism has been aimed at multinational tourism corporations in developing countries, including economic leakages, inappropriate forms and scale of tourism development, sunk costs and investment risks, over-dependence on multinationals and foreign domination. However, these corporations also bring investment funds, know-how, expertise, managerial competence, market penetration and control, and opportunities for local entrepreneurs (Lickorish 1991; Kusluvan and Karamustafa 2001). Based in Brazil, Atlantica Hotels is the largest multi-brand hotel administrator in South America, including 50 hotels in 30 Brazilian cities (Atlantica Hotels 2006). At a global scale, the InterContinental Hotel Group PLC owns, manages, leases and franchises through various subsidiaries over 3,650 hotels and 540,000 guest rooms in almost a hundred countries and territories. The company's portfolio includes InterContinental Hotels and Resorts, Crown Plaza Hotels and Resorts, Holiday Inn Hotels and Resorts, Holiday Inn Express, Staybridge Suites, Candlewood Suites and Hotel Indigo, and it has one of the largest loyalty programmes, Priority Club Rewards, with over 28 million members (InterContinental 2006). Lickorish (1991) argues that these corporations are bound by company law (national and international), by the interests and demands of their shareholders and always by the marketplace, which determines their profits. As investment opportunities may be less risky in developed countries, government agencies in developing countries often invite in these corporations and pay for joint scheme partnerships with private corporations.

Much of the attention in the tourism literature centres on multinationals; however, the small domestic firms and the informal sector in developing countries contribute a great deal to the industry. Gartner (2004) suggests that estimates of the number of small firms in many developing countries is far larger than the number of medium or large firms and accounts for between 40 and 90 per cent of non-government employment. 'Small firms which are the dominant form of tourism business at the destination level are the backbone of a destination's tourism economy' (Gartner 2004). In a study in Ghana, Gartner (2004) found that these firms face many challenges, not only from global factors and issues of dependency, but they also face risks from their inability to raise capital, lack of managerial skills, and their cultural obligations, which leads to the hiring of friends and family. In addition to small firms are the individuals who work in the informal sector, such as tour guides; these positions are very significant to the local economy in many destinations, though they are not accounted for in formal employment figures.

International agencies

International agencies also have a role to play in the tourism development process. Tourism consultants are often called upon to offer their expertise in the design and development of specific projects. Developing countries have been assisted by agencies such as the World Bank or United Nations agencies (Lickorish 1991). As illustrated in Box 4.2, the World Tourism Organization offers a tourism consultant service. In addition to planning expertise, international agencies have provided financial assistance. As indicated in Chapter 3, tourism is increasingly being written into Poverty Reduction Papers as part of loan applications to the IMF and the World Bank. In a study of World Bank involvement in tourism projects in the 1970s, Davis and Simmons (1982) identified 24 projects covering the Mediterranean basin, Mexico and the Caribbean, Africa and Asia. Involvement covered a range of elements including infrastructure, construction and rehabilitation, lines of credit and technical assistance. Hawkins and Mann (2007) update the role of the World Bank in tourism development to 2006 and note that there has been a shift to a more micro-level and policy intervention targeted at outcomes such as improving the livelihoods of local people. This shift is indicative of the recent focus on poverty reduction and the UN Millennium Development Goals. The Multilateral Investment Guarantee Agency (MIGA), which is part of the World Bank Group, is a global political risk insurer for private investors and companies that are investing outside their home country. MIGA guarantees investments against the risks of currency transfer, expropriation and war and civil disturbance. In Costa Rica, Conservation Ltd and the Bank of Nova Scotia, Canada went through MIGA to provide insurance for their ecotourism investment in the Rain Forest Aerial Tram that takes tourists on a 90-minute ride through the tree-tops in a rain forest near

Box 4.2

The UN World Tourism Organization and tourism planning consultants

As tourism continues to grow, the tourism consultant business also grows. If a government or business organization decides to pursue a tourism development project it can typically use its own staff members to prepare the designs and planning for the project or it can hire external consultants. These organizations provide a wide range of consulting services, from creating national tourism master plans to specific site plans with architectural designs. National governments can also obtain the services of tourism consultants through the United Nations World Tourism Organization (UNWTO) Technical Cooperation Service which was set up in 2004. The purpose of the Service is to promote and develop tourism in transitional and developing countries. At the request of a member government, the UNWTO Technical Cooperation Service hires individual consultants or larger consultancy firms to go to the destination and work on the technical missions and projects. Funding for these projects comes from a variety of agents including the United Nations Development Program (UNDP), the World Bank, the European Union, the Asian Development Bank, bilateral donors and others. The areas that the Service covers include:

- Identification and assessment of potential tourism development areas;
- Establishment of coherent frameworks for long-term sustainable tourism development;
- Preparation of national and regional Tourism Development Master Plans;
- Development of community-based tourism;
- Alleviation of poverty through tourism;
- Development of rural and ecotourism;
- Development of human resources for tourism;
- Formulation and implementation of appropriate marketing and promotional strategies;
- Strengthening of institutional capacities of national tourism administrations;
- Adjustment and improvements in existing tourism regulations in accordance with international standards;
- Stimulation and promotion of public–private partnerships;
- Establishment of hotel classification systems;
- Deployment of information technology in tourism.

Individuals and tourism consulting firms can apply on the UNWTO Technical Cooperation Service website to be included in the database as potential experts who the Service would call upon to carry out the work. Projects are carried out in specific countries as well as at regional scales. At the regional level, for example, work was conducted through the Service on the Silk Road Regional Programme (SRRP), which involves China, Kazakhstan, Kyrgyzstan, Tajikistan and Uzbekistan. Specifically, work was done to update a study on visa requirements for the Silk Road countries as well as to prepare a report on the inventory of the Silk Road tourism resources. Elsewhere, with the support of the European Union, the UNWTO Technical Cooperation Service is helping the South Pacific Tourism Organization (Fiji, New Guinea, Tonga, Tuvalu and Vanuatu) to develop a standardized system for recording, classifying and analysing tourism statistics. In Mali, Africa, the UNWTO Technical Cooperation Service sent a mission to prepare the guidelines to develop a future Tourism Master Plan. The mission evaluated the current state of the tourism sector and made many recommendations including promoting cultural tourism (The Grand Mosque at Djenne is a UNESCO World Heritage Site), as well as ecotourism and adventure tourism. Recommendations were also made on the development of individual tourism sites, administration of the tourism sector, accessibility, accommodation, environmental and economic impacts, and marketing. The framework has been approved by the government of Mali which is currently negotiating with the World Bank to get funding for the project to develop a future Tourism Master Plan. As with all plans, an important element is the need for monitoring and ongoing evaluation and adjustment. A final example comes from Yemen in the Middle East where the UNWTO undertook a mission to prepare an outline for a Tourism Development Strategy for the country and to review the institutional and legal frameworks for ecotourism on Socotra Island (known for its rich flora and marine wildlife). One of the recommendations was to strengthen the existing legal framework to protect the island from unsustainable tourism development and assist with licensing and regulation of ecotourism practices. One of the challenges with any report or plan which is generated is to what extent it will be implemented and later evaluated over time.

Source: UNWTO (2006).

San José (MIGA 2006). The challenge for developing countries which borrow money from international agencies is in considering to what degree they have to conform to the mandate of the funding agency. The funds may come with strings attached.

Regional-level organizations are also influential in tourism development. The Association of Caribbean States (ASC), which includes 25 Caribbean and Latin American countries, is taking an active role in tourism including improving air and sea transport infrastructure and services between member states (Timothy 2004). A special committee of the ACS was responsible for the Convention on the Sustainable Tourism Zone of the Caribbean (STZC). The ACS sustainable tourism plan includes a strategy to encourage community members and other stakeholders to participate in the tourism planning process, stimulate entrepreneurial activity, and to promote cooperation between government agencies and the private sector (Timothy 2004).

Not-for-profit organizations

Not-for-profit organizations or non-governmental organizations are attracting greater attention for the role they play in the tourism development process, and they range in scale from small local organizations to large-scale multinational NGOs. Their involvement may be in generating employment opportunities in community-based tourism initiatives, volunteer tourism, or pro-poor tourism. Volunteer heritage conservation groups have assisted in protecting sites as well as offering interpretation facilities to tourists. These organizations are also known for voicing concern over tourism development. Tourism Concern is based in the United Kingdom and is currently challenging Hilton for its claims of corporate social responsibilities (CSR). Conrad Hotels, the luxury brand of Hilton, has signed an agreement with a development company to manage Bimini Bay Resort and Casino in the Bahamas. The completed project will include a casino, condominiums, a golf-course and marina, and Tourism Concern argues that the development threatens a nearby fragile ecosystem (Tourism Concern 2006). The role of NGOs in community-based tourism is examined in greater detail in the next chapter.

Tourists

The final group listed in Figure 4.1 as an agent of development are the tourists. Tourists themselves have a role to play in the development process. Whether that is a more passive role, namely lying on the beach and then paying their hotel bill at the end of their vacation, or a more active role, namely participating in a volunteer tourism project helping to construct a local school or assisting on a farm, both can potentially contribute to the development goals of the destination. In addition, the actions they take, the behaviour they exhibit, their interactions with the locals and the environment can all potentially have an impact on tourism being a successful development tool (see Chapter 6).

Policy, planning and politics filter

Policy

Tourism development is about resource management, and Lickorish (1991: 160) argues that 'resource management starts with the preparation of an effective and realistic policy, and then a plan with related strategies for development and marketing which must be prepared together'. Politics and public policy is significant for tourism, whether it covers local or global scales, as politics regulates the tourism industry and tourist activity (Hall and Jenkins 2004). Hall (1994) argues that when governments adopt policies they are selecting from different sets of values which can have a direct impact on the form of tourism that is developed. The political ideology of a government can determine whether that government favours large-scale resorts or backpacker hostels, ecotourism or casinos (Elliott 1997). It is increasingly evident that many tourism policies are being set within the framework of sustainable development.

How specific policies come about, as Hall (1994) suggests, can arise within the context of a policy arena where interest groups (e.g. industry associations, conservation groups and community groups), institutions (e.g. government departments and agencies responsible for tourism), significant individuals (e.g. high-profile industry representatives), and institutional leadership (e.g. ministers of tourism and government officials) interact and compete in determining policy choices. According to Figure 4.1 the agents of tourism development such as a hotel developer will encounter what has been identified as the policy, planning and politics filter. Identified as a filter to represent the various layers of political, bureaucratic and regulatory administration, this filter is the screen that individual tourism developments must go through before they are constructed. The nature of this filter will be different in each destination, depending on the local political and administrative conditions. In some cases, local community groups or environmental groups may have the opportunity to voice their opinions during this filtering process, potentially altering the nature of the tourism development proposed. In the case of Mexico (see Box 4.1), the destination is very open to foreign investment; however, in the case of Vietnam, Lloyd (2004) found that there has been a turbulent relationship between foreign investors and the Vietnamese

Communist Party (VCP) who are governing a socialist market economy. In such an economy, there is tension between the state and its desire to retain a substantial role in the economy through control over privatization, monopolization of infrastructure and high-income-generating investments, and the pressure from foreign companies and lending agencies such as the IMF or World Bank that advocate for a reduced role for the state (Lloyd 2004).

Finally, an example of the impact of changes in ideology on policy and overall development was observed in Jamaica by Chambers and Airey (2001). During the 'Socialist Era' of 1972 to 1980 the Jamaican government pursued goals of self-reliance along with seeking to integrate tourism into the Jamaican way of life. In the second period of 1980 to 1989, 'The Period of Capitalism', the emphasis shifted to reducing government intervention and pursuing foreign exchange. During the first era there was some Jamaicanization of tourism and the policies contributed to a decline in arrivals, occupancy and hotel provision and employment. During the second era there was a recovery in tourism numbers but increasing tensions between locals and tourists. Government ideology then has the potential to impact upon tourism policy and development outcomes.

Types of planning

Once policies are established, then plans are written to ensure that developments reflect overriding policies. At a very broad level, Inskeep (1991: 25) defines planning as 'organising the future to achieve certain objectives' and it is carried out at different levels, from individuals planning everyday activities, to corporate planning, to governments creating formal comprehensive national or regional plans. A more specific definition linked to sustainability planning is a 'process which aims to anticipate, regulate and monitor change to contribute to the wider sustainability of the destination, and thereby enhance the tourist experience of the destination or place' (Page and Connell 2006: 477). Inskeep (1991: 25) identifies the major types of planning as economic development planning; physical land-use planning; infrastructure planning for services such as transportation, water, electrical, waste disposal and telecommunications; social facility planning for educational, medical and recreation facilities and services; park and conservation planning; corporate planning; and urban and regional planning. Liu and Wall (2006) argue that more attention needs to be given to planning tourism employment in developing countries so that locals have the necessary training and skills to participate in the benefits of tourism. Most of the major types of planning discussed above are completed by governments; however, corporate planning deals with the strategies that corporations put in place to enhance their business profile and generate profit. An airline will select which country to fly to or an international hotel company will select a domestic partner to construct the hotel that it will then manage.

Scales in planning

Tourism planning occurs across various scales and time frames. International organizations (e.g. the United Nations World Tourism Organization), regional trading blocks (e.g. Association of South East Asian Nations (ASEAN)) and international conservation and environmental laws (e.g. World Heritage Convention) all have a potential influence on tourism policy and planning (Hall 2000). With increased competition, tourism planning is occurring across borders and the importance of partnerships continues to be recognized as vital for tourism development. The Growth Triangle of Singapore, Indonesia and Malaysia is promoted as a site for investment for multinational corporations including those involved in tourism, and the Indonesian island of Bintan and Singapore are marketed together as tourism destinations (Timothy 2000). Individually, island states face a number of challenges including economic, social, institutional and environmental constraints, and often exhibit dualism whereby there is a large-scale technologically progressive export sector along with a small-scale, fragmented and undercapitalized domestic sector (McElroy and de Albuquerque 2002). Baud-Bovy and Lawson (1998) outline the main emphasis planning takes at national, regional and local levels. At the national level of tourism, master plans establish the broad framework for tourism creating environmental, economic and social policies. At the regional level, development strategies and structural plans for the region are established including a focus on regional infrastructure, protection areas and transport. At the local level, plans typically focus on local development, allocation of resources, conservation measures, zoning of land use, densities, coordination and implementation of policies. Finally, at the project level, the focus is on market and financial appraisal, organization of investments, site acquisition, facility planning and construction, and the coordination of development and operational needs (Baud-Bovy and Lawson 1998). In national or regional plans, time frames typically extend over a ten- to twelve-year period, with specific development programmes set out within the broader framework with shorter time frames, such as three to six years (Baud-Bovy and Lawson 1998).

In addition to scale, planning documents written for tourism have different purposes. Table 4.1 gives a list of various types of plans from Indonesia. The national tourism policy document sets out the framework for tourism in the country. More

Table 4.1 Examples of tourism plans in Indonesia

Type of plan	Title
National policy	Tourism Sector Programming and Policy Development, Output 1, National Tourism Strategy (UNDP 1992)
Island tourism	Tourism Development Plan for Lombok (JCP 1987)
Village tourism	Village Tourism Development Programme for Nusa Tenggara (WTO 1986)
Airport feasibility study	Feasibility Study for Airport Development in Lombok (Sofreavia *et al.* 1993)

specific plans include the development of tourism on the island of Lombok, to village tourism in Nusa Tenggara and a feasibility study for an airport. All are linked to tourism yet have different focuses. Community-based tourism is examined in Chapter 5. In Malta, the national tourism authority has just released its draft Strategic Plan for 2006 to 2009 written so as to be in line with the Malta Tourism Policy, the National Tourism Plan for the Maltese Islands, and the European Tourism Policy (Malta Tourism Authority 2006). Hall and Page (2006) highlight the importance of strategic planning in the context of sustainability. These documents outline a strategy which is a means to achieve a desired end. Strategic planning integrates planning and management in a single process and guides future direction, activities, programmes and actions on an ongoing basis. Strategic planning is meant to be iterative whereby planning systems adapt to change and learn, and are becoming increasingly important as a form of tourism planning.

Changing approaches to planning

Traditionally, tourism planning theory and practice has been associated with town planning (Costa 2006). The focus has been on land-use zoning or development planning at local or regional levels relating to site development, accommodation, building regulations, density of development, and presentation of cultural, historic and natural tourist features (Hall 2006). In recent years, governments have had to adapt to include concerns over environmental and social impacts of tourism and demands for more sustainable forms of tourism overall (see Chapter 2). These changes have occurred in an era when there have been demands for smaller governments (Hall 2000). Hall (2002) outlines the traditions of tourism planning by highlighting the work of Getz (1987). These traditions include boosterism; an economic, industry-oriented approach; a physical/spatial approach; and a community-oriented approach (see Chapter 5). In addition to these four traditions, Hall (2000) adds a sustainable approach to tourism planning. What these traditions mean is that those responsible for tourism planning would draft plans that have a specific focus as guided by the tradition under which they are operating. Both community and sustainable development approaches to planning tend to be more bottom-up planning, allowing for input from local residents as opposed to top-down planning where the planners are viewed as the experts. With shifts towards sustainability there are increased calls for incorporating locals into the planning and development process; however, locals are frequently under-represented as investors and decision makers, as they have a lack of knowledge about tourism and associated skills, and the priority is often placed upon economic growth by policy makers that has little concern for equity (Cohen (1982) in Liu and Wall 2006). Liu and Wall (2006) argue that if the locals are to really benefit and participate, efforts must be made to go beyond calls for more local involvement and move to incorporating human resource development into planning, thereby increasing the capabilities of the locals. If this is not done, benefits will continue to go to outsiders while the locals have to adjust to the changes that tourism brings (Liu and Wall 2006).

One of the challenges of tourism is that it is highly fragmented, which has caused problems for tourism planners and managers (Jamal and Getz 1995). Jamal and Getz propose a collaborative community-based planning process as a way forward to getting groups to work together. It facilitates public and private sector interactions, as well as providing a mechanism for community involvement in tourism. Similarly, Timothy (1998) also calls for tourism planning which requires cooperation between government agencies, between various administrative levels of government, between same-level autonomous polities, and between the private and the public sectors.

Costa (2006) suggests that new paradigms continue to emerge in tourism planning, and the current shift points towards a paradigm of tourism planning determined by market-led approaches. Tourism destinations will be planned and managed by models that place emphasis on the coordination and stimulation of private sector organizations that also bring public participation to the core of the decision-making process. Emerging approaches will give priority to improving coordination between private and public sector organizations and placing citizens (residents and visitors) at the core of decision making. Costa (2006) predicts that governance of tourism destinations, their capacity to be more sustainable, competitive and profitable, and the design of self-sustained development will become the priority of tourism planning models. Organizations will be set up around tourism clusters ('product space organizations') replacing old bureaucratic organizations ('space product organizations') and tourism administrations will adapt to this new trend of following clusters. Within these clusters, groups will have to work collaboratively to market their destinations. Jamal and Jamrozy (2006: 168) present an integrated destination management framework that brings in the destination's ecological-human communities as equitable and integrated members of planning marketing and goal setting. Within the framework they advocate sustainable tourism marketing and they argue that, based on an ecosystem network model, a sustainable planning-marketing orientation satisfies not only the needs and wants of individuals but also strives to sustain ecosystems.

With increased calls for tourism to be more sustainable, there is the recognition that tourism cannot be planned in isolation but rather needs to be integrated into broader development plans, and the public, private and not-for-profit sectors have to work together. This is particularly important if tourism is to be incorporated into broader development plans for the destination.

Regulatory environment

In addition to various policies and plans, any proposed tourism development also encounters the regulatory environment in the destination. The regulatory environment covers a wide range of specific regulations or laws such as building codes, labour laws, waste management regulations or environmental regulations that may be specific to the destination. For example, environmental impact assessments are typically required before development proceeds; however, different countries have different standards of regulations and degrees of enforcement. Some regulations will be strictly enforced and others will not, thereby possibly presenting some developers with the added advantage of choosing to locate their firm in one location over another. McElroy and de Albuquerque (2002) highlight Anguilla, Bermuda, St Lucia and the US Virgin Islands that have useful models of comprehensive development plans for small islands, and they note that most require a permitting process that often involves public hearings, environmental impact statements (EIAs) and social impact statements (SIAs), along with cultural and historical inventories. The Balearic Islands (Majorca, Minorca, Ibiza and Formentera) introduced an ecotax on tourists in 2001; however, with a change of government it was repealed in 2003. In Box 3.2, Wood (2004) indicates that cruise ships choose to fly flags of convenience in part to take advantage of reduced environmental and labour regulations. Different countries also have different regulations with respect to contesting planning decisions approved by the state. Developers may want to consider whether there are protests over a potential development, and if there is a due process for opponents to have their claims heard. Lengthy delays with respect to public hearing regulations may also cause developers to select another location.

Tourism is very much influenced by politics. Potential tourism development projects encounter policies, plans and regulations covering a variety of different scales. To these elements one must add the political situation in the destination and which individuals or groups hold the balance of power, not only in terms of formal but also informal power. Do local élites control the industry and the government, thereby dictating what forms of tourism are preferred and where they should be located? The heading of this section 'Policy, planning and politics filter' is meant to highlight the various complex layers through which a development has to go before any construction can begin.

Resulting tourism in the destination environment: form and function

Models of tourism development

Tourism development not only evolves from individual decisions about specific projects but also from within the context of multiple decisions, influencing over time the nature of development. There have been a number of models presented in the literature that examine the changes in resort morphology in destinations over time. Miossec (1976), cited in Opperman and Chon (1997), developed a model that looks at the evolution of resorts, transport routes, tourists and hosts. In the final stage of the model, there is a hierarchy of resorts that are highly connected and there are also excursion routes into the interior of the hypothetical island. As destinations open up, tourists have the potential to travel greater distances not previously accessible, thus increasing level of contact with locals. Not only does improved infrastructure for tourism promote better access to the destination for tourists, it also induces migration of people looking for work in tourism. This migration often leads to socio-cultural, economic and ecological change, as has been found in Zanzibar, Tanzania (Gössling and Schultz 2005) and Goa, India (Noronha *et al.* 2002). One of the most enduring models is the Tourism Area Cycle of Evolution, which is linked to the concept of a product life cycle. The model, developed by Butler (1980), suggests that tourism developments go through the stages of exploration, involvement, development, consolidation, stagnation, and decline or rejuvenation. In this model, as time passes, control of the industry tends to move from local control to more external control as large-scale multinational tourism companies open up resorts once a destination achieves more recognition. If a destination goes into decline, decisions need to be made as to what strategy should be used to rejuvenate the area. While some resorts have tended to follow the stages in the model, other destinations have jumped several stages, resulting in more instantaneous development rather than development that occurs over a longer period. As in the case of Mexico, the government has purposely selected undeveloped areas and built large-scale resorts which are set up as regional growth-poles (Telfer 2002b).

Wall (1993) developed a tourism typology consisting of attraction types (cultural, natural and recreational), location (water or land based), spatial characteristics (nodal, linear and extensive), and development strategies (highly developed, developed and developing). Accommodation type is regarded as a key element in the tourism system with implications for the characteristics of tourists, the built environment, economic impacts, degree of local involvement, and critical environmental and sustainability factors such as capital, land, water, energy and waste disposal systems. Wall (1993) suggests that a mix of tourist types (mass to explorer) and accommodation types (five-star hotels to guesthouses) can be integrated to promote sustainable development. In developing considerations for appropriate and sustainable tourism development, Telfer (2002a) examined (1) the scale and control of development, and (2) local community and environmental linkages. Under each of these main categories, the following subcategories were examined:

Scale and control of development

- Focus of development, scale, rate of development, level of economic distribution, type of planning, local involvement, ownership, industry control, role of government, management origin, accommodation type, spatial distribution, tourist type, marketing target, employment type, infrastructure levels, capital inputs and technology transfer.

Local community and environmental linkages

- resource use, environmental protection, hinterland integration, intersectoral integration, cultural awareness, institutional development and local compatibility.

An investigation of selected tourism development strategies

The selection of any specific tourism development strategy needs to be understood within the framework given in Figure 4.1 and who is in control of the development process. For example, is the strategy for development based on a government tourism master plan or is it being developed by a private entrepreneur as part of a broader investment strategy in tourism to satisfy shareholders? Private investment may or may not coincide with national, regional or local tourism plans, and various levels of government may have competing agendas. As Reid (2003: 225) suggests, 'local and regional development requires different processes and foci than does the goal of raising foreign currency in order to pay off foreign-held debt'. The options for development are as diverse as the number of different types of tourism; however, the selected strategy will also be governed by the politics of place and resources in the destination (natural, human, economic, infrastructure, technological). Some destinations will have an advantage in resort development, while others have, over time, been branded either formally or informally as ecotourism destinations. As suggested in Figure 4.1, the resulting tourism in a destination obviously has form but it also has function linked to the development process. From large scale to small scale and from enclave resorts to integrated resorts, the degree to which tourism establishes links to its destination will have a major bearing on tourism contributing towards the specified development goals. Those types of developments that are strongly integrated with local, regional or national economies will generate greater potential for more people residing in the destination to participate. However, those developments that are primarily linked to international rather than domestic economies will be more removed economically from the local population and may locate the destination in a very isolated position. It is important to note, however, that even these types of development can be strong generators of tourist numbers and foreign currency and, in many cases, are the best option for developing countries that initially do not have the capacity to effectively launch their own domestic tourism industry. While it is not possible to examine all types of tourism within the context of this chapter, this section will focus on a selected number of tourism developments and point out the strengths and weaknesses of each type in terms of their potential to contribute to development.

Resort development

Large-scale resorts targeting mass tourism often represent the main development strategy in many developing countries as they have the advantage of attracting a significant number of tourists, income and foreign investment (see Plates 4.1, 4.2 and 4.3). Box 4.1, on Mexico, clearly illustrates the economic potential of large-scale beach resort development and so this development option cannot be ignored. An example from Brazil highlights both domestic and international involvement in beach resort development. The Brazilian government, working to update the tourism sector, borrowed US$800 million from the Inter-American Development Bank to improve infrastructure in the northeast and created the Ministry of Tourism in 2003. The country is becoming a destination for European tour operators looking for something different from the Caribbean. Spanish and French multinational hotel companies have also been opening up resorts in the country, and there has been domestic investment in resort development. In 2000, Previ, a pension fund for employees at Banco Brasil (the country's largest bank), partnered

Plate 4.1 ***Cuba, Varadero: Beach resort.***

Plate 4.2 ***South Africa, Sun City Resort: Tourists swimming at the man-made beach at the resort in a water-scarce area.***

Plate 4.3 ***Tunisia, Monastir: Luxury beach resort, Amir Palace Hotel. Note the architectural design.***

with Odebrecht, a large Brazilian construction firm, to spend US$200 million to build five resorts, six smaller inns, conventions and sports centres, restaurants, stores, swimming pools, tennis courts and an 18-hole golf course on the Sauipe coast. Previ later bought out Odebrecht's stake in the project and then leased the five resort hotels to multinationals. Two were leased to the US chain Marriot, two were leased to France's Accor and one to Jamaica's SuperClubs (Kepp 2005). Oppermann and Chon (1997) point out that in many developing countries prosperous hotel chains such as Atlantica Hotels from Brazil, which is the largest multi-brand hotel administrator in South America, have emerged. While some argue that sun-and-sand tourism is entering a decline as a result of environmental damage and a shift in consumer demand to post-tourists or 'new tourists', Aguiló *et al.* (2005) argue that is not the case in the Balearic Islands. Following restructuring and quality improvements, it is still a competitive destination. The authors, however, argue that the sun-and-sand model of tourism needs to be adapted to the framework of sustainable development.

While these resorts bring in large numbers of tourists, there is lively debate over their overall benefit to the destination. Criticism has been aimed at these resorts, especially if they are enclave resorts cut off from the local community. They are often controlled from abroad, generate high rates of leakages, and the multinationals are more interested in profit than in establishing strong links with the local community. In Cuba, locals are allowed only reduced access to Varadero beach hotels. As noted in Chapter 3, Britton (1982) developed an enclave model of tourism. Britton is especially critical of package tours. Tourists are transported directly to resort complexes where they typically reside for the duration of their trip. Travel in the destination is only between resort clusters and to the airport to return home. The transportation, organization of tours and accommodation occur largely within the formal sector. Criticisms have also been directed to the trend of all-inclusive packages at these resorts, which include transportation, accommodation and meals. Following arrival at the resort, tourists need to be enticed to venture out of the complex. All meals will have been paid for in advance, so tourists may be hesitant to venture out to a local restaurant to pay for another meal, thereby reducing the potential multiplier effect. Tips are often included in the package, thus removing this opportunity for hotel staff. In an examination of three resort enclaves in Indonesia, including Nusa Dua on Bali, Kuta on Lombok and Bintan Beach Resort on Bintan, Shaw and Shaw (1999) argue that the notion of enclavity is inherently unsustainable, marginalizes local entrepreneurs, and widens the social, economic and cultural gaps that already exist between hosts and guests. Similarly, in the Okavango Delta of Botswana, Mbaiwa (2005) found that enclave tourism was dominated by foreign companies. While tourism here is in much smaller numbers since they are following a policy of high cost and low volume for safari tours, concerns over enclave developments have surfaced. This type of development has led to the repatriation of tourism revenue, domination of management positions by expatriates, and lower salaries for citizens. Mbaiwa (2005) argues that tourism has failed to contribute significantly to rural poverty alleviation in the Okavango region.

Integrative development and alternative tourism

While criticisms have been aimed at large-scale resorts and enclave resorts, others have called for more integrative tourism development that promotes linkages to local communities. At one end of the spectrum of integrated tourism are larger tourism developments and at the other end are small-scale or alternative tourism destinations where locals are given more opportunity to participate. Telfer (1996) and Telfer and Wall (1996, 2000) examined the efforts of both large- and small-scale hotels in Lombok and Yogyakarta, Indonesia to establish linkages to the local agricultural sector (see Plates 4.4 and 4.5). In both locations, hotels established direct and indirect links to local suppliers, traditional markets and farmers for the purchase of food products. While these initiatives are promising and generate increased multipliers, there are challenges to maintaining these relationships long term, especially when five-star hotels place increasing demands on small producers and suppliers at certain times of the year. Another example of alternative tourism which involves locals directly is village-based tourism. In the case of the Solomon Islands, village tourism takes the form of tourists travelling by motorized canoe between villages situated on the edges of lagoons (Lipscomb 1998). Tourists undertake bush walking, snorkelling, historical and nature tours, cultural displays, and arts and crafts in each village, as well as staying overnight in the village to get a greater understanding of village life. Lipscomb (1998) suggests

Plate 4.4 ***Indonesia, Lombok: Fisherman-turned-supplier in black leather jacket purchases fish in a local fish market which in turn will be sold to an international hotel (see Telfer and Wall 1996).***

that the potential for village-based tourism to persist at an early or exploration stage of development for a long period of time, producing a small income for the village, is in itself a compelling argument to develop village tourism. There are, however, challenges with this type of tourism including marketing, planning and cultural problems, and the fact that there is a danger that the local élites may be primarily the ones who benefit. The emphasis on integrated or community-based tourism as a form of sustainable tourism, where the control and benefits remain in the local community including pro-poor tourism designed specifically to address poverty issues by enhancing opportunities for those in greatest need, will be addressed further in Chapter 5.

Ecotourism

Ecotourism continues to receive attention, since theoretically it represents a win-win scenario through environmental protection and improved local livelihoods (Cater 2004). For many developing countries it is promoted as a means of reconciling economic growth and environmentally sustainable development (Duffy 2006). It is often seen at the opposite end of the spectrum from mass tourism and part of alternative tourism. Ecotourism strategies have been endorsed by tourism businesses and destinations but are also advocated and promoted by a wide range of agencies such as the World Tourism Organization, United Nations Environment

Plate 4.5 ***Indonesia, Lombok: Small local fruit and vegetable supplier makes a delivery to the Sheraton Hotel in Sengiggi Beach. The products were purchased at local markets.***

Programme (UNEP), international lending agencies (e.g. World Bank), indigenous rights groups, development non-governmental organizations (NGOs), environmental NGOs (e.g. World Wildlife Fund and Conservation International), the International Ecotourism Society, and bilateral development agencies (Cater 2004; Duffy 2006; Jamal *et al.* 2006). While there have been debates as to precise definitions of ecotourism, Diamantis (2004: 5) puts forward as a guiding conceptual principal that ecotourism 'occurs in natural settings (protected and non-protected) with an attempt to increase benefits to the economy, society and environment through sustainable educational practices from locals to tourists and vice versa'. It is particularly promoted in the undeveloped world, as it is their 'underdevelopment' or 'lack of modernization' which makes these environments attractive to tourists from the 'developed' world (Duffy 2006). Ecotourism is promoted as a means for poorer communities to generate income, and for many communities living adjacent to national parks or reserves it is presented as a beneficial return 'for relinquishing rights over using the plant and animal resources within those reserves for subsistence purposes' (Duffy 2006). The United Nations declared 2002 as the International Year of Ecotourism. During that year the World Ecotourism Summit was held and resulted in the Quebec Declaration on Ecotourism. The Declaration recognizes that ecotourism embraces the principles of sustainable tourism with respect to the social and environmental impacts of tourism, while espousing the following concepts that distinguish it from the wider issue of sustainable development (UNEP/WTO 2002 in Cater 2004: 485):

- It contributes actively to the conservation of natural and cultural heritage;
- It includes local and indigenous communities in its planning, development and operation, and contributes to their well-being;
- It interprets the natural and cultural heritage of the destination for visitors;
- It lends itself better to independent travellers, as well as to organized tours for small groups.

Duffy (2006: 2) suggests that the politics of ecotourism is revealed in debates over definition and raises the question: Can it be 'provided by global tour operators and luxury nature based resorts, or is genuine ecotourism found in small scale local community run projects and campsites'? Efforts have also been made to develop certification programmes, and one of note from a developing country is the Costa Rican Sustainable Tourism Certification (Jamal *et al.* 2006). Weaver (2004) outlines the structural dimensions of ecotourism, which range from hard to soft ecotourism. Hard ecotourism traces its origins to the 1980s as a form of nature-based, small-scale alternative tourism that emerged as a reaction against the perceived environmental, socio-cultural and economic excesses of large-scale development. Traits of hard ecotourism include a high level of environmental commitment among ecotourist participants involved in specialized, often physically and mentally challenging experiences over longer periods of time and in smaller numbers. At the other end of the spectrum is soft ecotourism, which has traits of a moderate or 'veneer' commitment to environmental issues and participants are conventional tourists experiencing ecotourism in larger numbers as one part of a diversified experience. They have extensive reliance on services and an emphasis on package travel through travel agencies and tour operators. A cruise ship, for example, may offer a tour to a nature garden or reserve. Weaver (2004) points out that while hard ecotourism purists would not agree with soft ecotourism, it is perhaps soft ecotourism that can generate the financial objectives to help enhance the natural environment and to be managed into site-hardened intensive zones.

The diversity of ecotourism is evident in South Africa where there are national parks with no accommodation facilities (e.g. Knysna National Park) while others, such as Kruger National Park, have luxury accommodation, extensive road networks, four-by-four routes, wilderness trails, guided safari drives, swimming pools, golf-courses and banking facilities. In South Africa, steps have been taken to commercialize the ecotourism industry. Many protected areas have management plans that employ zoning to protect some areas while allowing tourism to be developed in other areas. The commercialization of Kruger National Park is now being undertaken by the private sector, with companies submitting proposals which are evaluated in terms of financial aspects, environmental management, social objectives and empowerment. The linking of tourism with environmental protection and empowerment for human development illustrates the potential benefits that can occur if nature-based tourism is properly planned and managed (Spenceley 2004). Elsewhere, ecotoursim lodges, such as those investigated by Johansson and Diamantis (2004) in Thailand and Kenya, demonstrate how the private sector can also contribute towards providing benefits to host communities.

While the various developed forms of ecotourism continue to be a source for debate, the focus on the contribution to conservation and the involvement with local communities is not always followed. As Cater (2004) points out, the ideals of ecotourism have met with harsh market realities, resulting in a considerable divergence between theory and practice, and ecotourism has frequently been misinterpreted, misappropriated and misdirected. Products are marketed as ecotourism when they do not meet the basic criteria of the concept. Writing ten years earlier, Cater (1994) warned that ecotourism may share many of the same characteristics as mass tourism in terms of leakages, since much of the tourist expenditure is not made in the destination. Citing the case of Belize, Cater (1994) notes that there is a high degree of foreign investment in ecotourism which has created inflationary pressures in the local economy. As sites become popular, the concentration of visitors leads to degradation, and there is debate about whether ecotourists themselves are 'an environmentally sensitive breed' (Cater 1994: 76).

Duffy (2006) raises a number of important issues in examining the politics of ecotourism in developing countries. In addition to issues of definition, there are concerns over how destinations are marketed as undeveloped regions perpetuating a particular image of the developing world. Ecotourism resorts also rely on global networks of travel and global tour operators, raising issues of sustainability and power and control. Power issues are also evident in the fact that funding from international agencies may also come with strings attached and, at a local level, not all residents will benefit equally from an ecotourism resort (Duffy 2006). Duffy (2006) argues that ecotourism's links to the promotion of neoliberalism are the path to development. Jamal *et al.* (2006) argue that ecotourism has been developed along a modernistic and commodified paradigm and it needs to be reoriented to a social-cultural paradigm based on participatory democracy and meaningful relationships with the biophysical world. Community participation in tourism will be explored further in Chapter 5.

A final trade-off considered here with respect to ecotourism is the financial aspect. While this sector is meant to protect and enhance the natural environment, it also dictates that lower numbers be attracted to the destination in order to protect the initial resource sought out by tourists. Lower numbers, unless they are high-paying tourists, result in smaller overall profits at the destination level compared to more conventional forms of tourism. On an individual basis, those directly involved in ecotourism will see an improvement in income; however, it may be difficult to gauge how this contributes to overall levels of development for the entire destination. If the destination becomes well known, not only tourists but also those seeking a job in the industry will visit the destination, causing a variety of impacts if not properly controlled.

Culture- and heritage-based tourism

Cultural attractions and historic monuments are part of the attraction in many developing countries (see Plate 4.6). Cultural attractions generate tourism flows

Plate 4.6 ***Thailand, Bangkok: Tourists visiting the Grand Palace Complex.***

and thereby present opportunities for locals to interact with and generate income from the tourists. Festivals, souvenirs, traditional dances and local food are all part of the attraction. Smith (2007) suggests that cultural tourism is changing and, while there will still be opportunities to enjoy unique cultural experiences, cultural excitement is more likely to exist in cosmopolitan locations (e.g. world cities) than in small villages. Whether cultures are associated with a state, a region or a specific ethnic group it is important to consider how those cultures are portrayed, who has ownership over how the culture is presented, and whether or not locals are being exploited as culture is incorporated into tourism. Wood (1997) explores the relationship between the state, identity and tourism, and notes two examples of the state having a major influence in culturally related tourism. In China the government officially recognizes 55 ethnic minority 'nationalities' that mostly reside in officially designated autonomous regions and these range in size from a village to a province. The Chinese government designates official tourist sites and then determines which areas should be opened up to foreigners. In Singapore the state and the tourism industry treat the country as having four categories of ethnicity (Chinese, Malay, Indian and Other). Wood (1997) comments that these four categories do not accord with the self-identity and lived experiences of Singaporeans; nor do they reflect ethnic tradition. However, reinforcing these labels for the tourism industry raises questions of authenticity.

Authenticity has been widely discussed in the tourism literature (Hashimoto and Telfer 2007). In the context of tourism 'authentic' has been used to describe products such as works of art, cuisine, dress, language, festivals, rituals, architecture, or anything that is part of a country's culture (Sharpley 1994). However, as Timothy (2005) suggests, authenticity is a subjective notion that can vary from place to place, culture to culture and from person to person. Hughes (1995) illustrates political influence as follows: 'authenticity in tourism is held to have been produced by a variety of entrepreneurs, marketing agents, interpretive guides, animators, institutional mediators and the like' (Hughes 1995: 781). Tourist behaviour will be explored in greater detail in Chapter 6; however, tourists do have different motivations and some are more concerned with experiencing the 'authentic' cultures than others.

Wall and Xie (2005) examined the Li dancers of Hainan, China who are members of an ethnic minority employed to represent their culture to tourists. The following study focuses on authenticity and uses five themes, which for the purpose of this chapter can be broadened to reflect on the role of cultural tourism and development. The five themes or continua are:

- Spontaneity versus commercialism;
- Economic development versus cultural preservation;
- Cultural evolution versus museumification (the freezing of culture);
- Ethnic autonomy versus state regulations;
- Mass tourism versus sustainable cultural development.

The authors found that there is a desire to celebrate and portray ethnicity through commodification and that mass tourism is desirable, as it leads to job security and hence economic prosperity. The commodification of culture is viewed as a positive mechanism in the pursuit of sustainable development, as it is seen as inseparable from economic development, which is desirable. However, the authors do raise a cautionary note that the control of the commodification of culture and the associated benefits are not in the hands of the minority. While cultures tend to evolve with or without tourism, how culture is used and who benefits will influence the overall contribution to development broadly defined. A further discussion of the social and cultural impacts of tourism is found in Chapter 7.

Living cultures are often linked to historic monuments, which are increasingly being used as tourist attractions (Timothy and Boyd 2003). UNESCO designates World Heritage Sites for countries that are participating members (see Plate 4.7). This presents both opportunities and challenges for local communities. The designation may generate additional tourists but there is also the danger of attracting too many, causing problems for sustainable development as well as raising questions of control over the site. Evans (2005) examined the promotion of Mundo Maya linking the heritage sites of pre-Columbian civilizations with the all-inclusive Mayan Riveria resorts of Cancún and Cozumel in Mexico. Evans (2005) explored the power relations between the indigenous communities, the state and dominant groups, along with the marketing of the region. He found that there is an absence of genuine local community and cultural involvement in the heritage site management, and in some cases interpretation, all leading to a pattern of commodification. Evans (2005: 46) states that this is perpetuating a 'system in which national authorities and power groups (including heritage intermediaries)

Plate 4.7 ***Tunisia, El Jem: Roman Colosseum. UNESCO World Heritage Site.***

effectively collude in international agency and corporate foundation programmes and development aid, using the dualistic heritage tourism and conservation rationale for intervention'. As with any tourism development, steps need to be taken so that locals are incorporated into the planning of the site so that they do not become marginalized and only serve to be part of the tourism product and not the beneficiaries.

Tourism development outcomes: the need for increased linkages and participation

The final element in Figure 4.1 is the development outcomes as a result of tourism. What then can tourism do to contribute towards social, economic and environmental goals of development? Which of the above tourism options could a destination select as a viable development? Page and Connell suggest that tourism development has the potential to assist with development in terms of poverty reduction and employment growth. However, if tourism is viewed from a broader political dimension then tourism can transform an economy in the following ways (Page and Connell 2006). Tourism can be a way of obtaining hard currency to improve the balance of payments and indebtedness. It can be a catalyst for social change, establishing greater contact between the indigenous community and the tourists. Tourism can also be seen as a symbol of freedom, whereby citizens are allowed to travel freely within and outside their country. The industry can be a mechanism for improving local infrastructure to serve both tourists and locals. Tourism can be an integral part of economic restructuring through privatization, as well as exposure to transnational corporations and national and international markets. Finally, tourism can complement commercial development through the growth of business tourism and by encouraging small-scale entrepreneurial activity (Page and Connell 2006). If planned and managed correctly, tourism can also contribute to sustainable development, underlying the need to preserve and protect resources for future generations.

What Page and Connell (2006) have outlined is perhaps an ideal that a developing country can aim for. Tourism development also comes at a cost, and so, in reality, achieving these goals is much more difficult. This chapter has outlined the tourism development process. The agents of tourism development, including the state, private corporations (both domestic and multinational), not-for-profit organizations, and related financial and technical planning organizations, as well as tourists themselves, all have their own values, ideology, goals, priorities, resources and strategy for tourism development. The state provides the framework for tourism development to occur in the destination by setting policy, creating plans for the destination and outlining any specific regulations governing tourism development. The policy and planning environment is a very political environment where the values of decision makers will greatly influence the type of tourism that actually goes forward for construction. Will the state support casinos or beach resorts or ecotourism? Will it choose to diversify and build tourism based on a number of different product options? The political structure in the destination, and to what degree the state can act in a more unilateral fashion or whether there is a more democratic process involved allowing various groups and individual citizens to voice their support or opposition, will also influence development. Tourism plans created by the state cover different geographic scales (local to national) to guide development; however, there can be competing interests between various levels of the state as well as between the state and private interests. As Reid (2003: 134) suggests,

> often, local communities are assumed to be represented by their local governments, but, as we have seen, national governments have their own agenda, usually having to do with earning foreign exchange in order to pay down debt held by foreign banks, but not necessarily related to issues like regional development and income generation at the local level.

In this age of globalization, private corporations will look for the destination that offers the most support and incentives and the fewest barriers. Tourism firms will conduct their own corporate planning and will, in most cases, locate their product where they can get the best value for their investment dollar.

Once a tourism project winds its way through the filter of policy, planning and politics, construction occurs (large scale or small scale, enclave or integrated) and its potential to contribute to development is largely based on the level of linkages to the destination economy. The greater the development project is linked to the local economy through purchasing products locally, hiring local people, using local services, involving area citizens in the planning process, and reinvesting in the area through infrastructure, the greater the net benefit. Part of the challenge is to what extent the local economy can be integrated into the tourism sector and to what degree locals can participate. If local products cannot be obtained in sufficient quality and quantity the challenges become greater. If, however, there is potential for increased backward linkages, efforts need to be made (or in some cases regulated) to enhance these partnerships. In the case of the hotel industry in China, Zhang *et al.* (2005) identified that there is great potential for outsourcing, yet the market is still immature due to structural barriers in place. Referring to a study of hotel development in Shanghai, the authors indicated that outsourcing was occurring in the areas of a shopping arcade, recreation centre, flower kiosk, beauty salon, restaurants, dance-hall, karaoke parlour, sauna, as well as some housekeeping and gardening departments. If the local community is able to successfully fulfil the outsourcing contracts, the backward linkages will continue to be enhanced. As economic growth from tourism occurs, there is potential for it to be funnelled into other areas of the destination society to help promote the broader social and environmental developmental goals. If, however, there are limited connections to the local economy and profits are repatriated out of the country, there will be limited potential for tourism to contribute to the development process. Telfer and Wall (1996) summarized from the literature the difficulties in locally sourcing agricultural products in developing countries for hotels. The main challenges with supply were with respect to quality and quantity of product, yet once linkages are established the other challenge becomes making these partnerships last. Other barriers, such as tourism being controlled by the local élite or the government using the funds from tourism to service international debt, will also have a negative effect on development outcomes.

Not only do the economic linkages need to be made; there is also a need for greater involvement of locals in the tourism planning process. However, as identified by authors such as Liu and Wall (2006), without adequate human resource management locals may not have the skills and knowledge to participate. They also suggest that 'many, perhaps most communities in the developing world may require an outside catalyst to stimulate interest in tourism development and external expertise to take full advantage of their opportunities' (Liu and Wall 2006: 160).

Planning and regulating for development

As illustrated in the previous chapter, there are competing arguments over the role of the state in the economy and planning. Through setting policies, plans and regulations, the state can direct tourism to promote development. The direction of tourism planning has changed over time and sustainable development has come to the foreground. Sustainability, competitiveness, profitability, self-sustained development, governance, public–private coordination, partnerships and local involvement are becoming priorities in tourism planning (Costa 2006). The strategies selected to develop tourism can range from facilitating the development of large-scale growth-poles to encouraging local entrepreneurial development or placing restrictions on development in certain areas to protect fragile resources. Tourism can be developed as part of a regional strategy or to revitalize an urban area. Hall and Jenkins (1998) outline a range of tourism development policy instruments for rural tourism; however, as illustrated through some of the examples below, they can be adapted to other locations and scales. The first category is regulatory instruments, which involve law, regulations, permits and licences. Laws, for example, can be passed to zone an area suitable for cruise ship development or a resort complex. The second category is voluntary instruments, and these include providing technical assistance, or assisting volunteer associations and non-governmental organizations. Governments may support a regional tourist organization or a heritage conservation group. The third category is expenditure, which includes spending money on specific activities, operating a public enterprise, public–private partnerships, monitoring, and evaluation and promotion. A government may have to spend money on improving an airport or want to enter into an agreement with a private company to develop tourism in a certain area. The fourth category is financial incentives including pricing, taxes and charges, grants and loans, subsidies and tax incentives, rebates and rewards and vouchers. As illustrated in Box 4.1, Mexico offers a variety of incentives to attract tourism developers. Economic incentives will attract developers. The drawback, however, is potentially giving up too much. The final category is non-intervention whereby a state may choose not to get involved in order to achieve its objective.

Within the options open to the state, regulating practices to promote backward economic linkages and improving or overcoming obstacles for local participation in the planning process are both options to promote development. Integrating tourism development with the local economy not only stimulates the formal tourism sector but also the informal sector which in some developing countries may be very important to overall development. Similarly, establishing programmes to assist local entrepreneurs will help enhance the potential for them to participate in the economy. By providing opportunities for locals to participate in the planning process, new opportunities may open up, creating the prospect that more locals may participate in the benefits of tourism. The inclusion of the community in tourism will be further explored in the next chapter. The decision as to how much regulation to enforce is difficult; investment incentives may attract developers, while regulations may drive them away. Who ultimately controls the industry will determine, to a large extent, where the benefits go. As illustrated in the examples in this chapter, there is no one form (or forms) of tourism development that is suitable to all destinations and no one set of plans, policies and instruments that can ensure success. It is also important to recognize that even with good plans there are often problems in implementation (Lai *et al.* 2006). There is increasing awareness, however, that tourism cannot be planned in isolation but needs to be integrated as part of broader development strategies within the context of sustainable development.

Discussion questions

1. **How does government policy impact upon tourism development?**
2. **What are the advantages and disadvantages of different types of tourism development?**
3. **What planning approaches can be used to maximize benefits to the local population?**
4. **What are the difficulties in implementing a sustainable development policy for a tourism corporation?**

Further reading

Hall, C. M. (1966) *Tourism and Politics: Policy, Power and Place*, Chichester: John Wiley & Sons. [This book provides a good overview of the relationship between tourism and the political process. It also covers topics such as international relations, political stability and culture.]

Hall. C. M. (2000) *Tourism Planning: Policies, Processes and Relationships*, London: Prentice Hall. [This book provides a good overview of tourism planning and the connection to sustainability. It explores planning at various geographic levels as well as the importance of cooperative structures for planning.]

Websites

The website for Fonatur in Mexico outlines its various tourism development projects along with investment incentives: www.fonatur.gob.mx/_Ingles/index.html.

The website for Wimberly, Allison, Tong and Goo, an international architecture, design, planning and engineering company with a focus on tourism. The web page has numerous case studies of tourism developments with which the company has been involved: www.watg.com.

The website for the Barbados Ministry of Tourism links to the document *National Development Through Sustainable Tourism* as well as to Barbados Tourism Investment Inc, the company with which the government is operating to attract investment. Investment opportunities are displayed in the web page: www.barmot.gov.bb/about.htm.

5 Community response to tourism

Learning objectives

When you have finished reading this chapter, you should be able to:

- **Understand the nature of communities;**
- **Be aware of how communities and individuals respond to the introduction of tourism development;**
- **Be familiar with the links between community involvement and sustainable development;**
- **Evaluate different community-based tourism initiatives.**

There is a wide range of perspectives that can be taken on communities in the context of tourism. For some, the community may be considered as the main attraction and the gatekeeper to local knowledge, while for others the community is simply the setting where tourism occurs (Mowforth and Munt 1998). For others still, a community may, in fact, stand in the way of other potential tourism developments and should therefore be moved. Communities are increasingly being drawn into tourism not only from the demand side, as tourists actively seek out new destinations and communities to experience, but also from the supply side, as communities are becoming aware of the potential of the products they can offer to tourists and the economic gains that can be made. An important question to consider is who controls community-based tourism and whether the benefits from tourism go to the local people or whether they are controlled by the local élite or external tourism development agents exploiting the local community. Communities are not homogeneous, and not all residents support the integration into tourism. Locals not only have to deal with the constant attention of tourists but also the potential commodificaiton of their culture (Boissevan 1996). With the shift towards sustainable development, there are increased calls for tourism to contribute in a positive way to the host community. Not only should tourism development protect and enhance local cultural heritage and environment; local residents also need to have input in the tourism planning process and greater participation in the benefits of tourism. With greater local control and by integrating tourism into the local economy through the use of local labour, products and resources, there is enhanced potential for tourism to contribute to the broader notions of development such as empowerment and greater self-reliance. In the context of indigenous peoples, Ryan (2005: 4) argues that:

> tourism is increasingly viewed not simply as a force for the creation of a stereotypical image of marginalised people, but a means by which these peoples aspire to economic and political power for self advancement, and as a place of dialogue between and within worldviews.

The purpose of this chapter is to explore the various relationships between tourism and local communities, focusing on how communities respond to tourism and some of the various concepts, programmes and agents that may promote development though the use of tourism. After examining the nature of communities and how communities respond to tourism, the chapter then focuses on the linkages between sustainable development and community-based tourism. Related issues such as participation and empowerment are considered. The power of communities in the face of tourism will be explored with examples ranging from relocation to resistance. Finally, the chapter will focus on several current issues in tourism as they relate to communities, including the role of NGOs, fair trade and tourism, pro-poor tourism, volunteer tourism, and gender and community development. While all of these initiatives hold the promise of increased opportunities and benefits for local communities, they face numerous difficulties and challenges which are explored throughout the chapter.

Nature of communities

For some tourists, experiencing the local culture, language, traditions, lifestyles and natural environment are essential components of the trip. Increasingly these elements are packaged and sold both by local inhabitants and by external agents under various labels such as cultural tourism, heritage tourism, indigenous tourism, village tourism and community-based ecotourism. This process is often referred to as commodification. However, debates continue as to whether the changes afforded by this type of tourism will forever change the host community in a negative way or whether in fact it may revive it. Modern advances in travel and the impacts of globalization have opened up distant communities to outsiders, which in the past may have had little contact with the 'outside world'. Harrison and Price go so far as to argue that no community in the world now exists in isolation and few ever did. 'Willingly or not, all communities are part of nation-states with policies for all aspects for their inhabitants' lives; for education, health care, communications and security (social and otherwise)' (Harrison and Price 1996: 2). The implementation of these policies in part establishes the framework for tourism (Harrison and Price 1996).

If communities in developing countries are increasingly becoming the focus of attention for tourism, what are the central attributes of the term 'community'? The concept of community is complex, and there have been various ways that communities have been contextualized. Rothman *et al.* (1995) conceive of community as the territorial organization of people, goods and services, and commitments that are important subsystems of society where locally relevant functions occur. Urry (1995) built upon the work of Bell and Newby (1976) who identified three main concepts of community. The first is a topographical sense, which can refer to the boundaries of a community. The second is sense of community as a social system implying a degree of local social interconnection of local people and institutions. The third is a sense of 'communion', a human association implying personal ties and a sense of belonging and warmth. To these three concepts Urry (1995) added the concept of ideology, which can often hide the power relations that inevitably underlie communities (Richards and Hall 2000a). With community linked to a geographic area, it is interesting to consider the diversity of destinations visited ranging from the rural Hill Tribe People in northern Thailand to the recent and controversial urban slum tours of Kibera (the largest slum in Africa near Nairobi, Kenya).

Richards and Hall (2000a) state that community has permeated the sustainability literature, and there are few sustainable tourism policies that do not refer to the importance of long-term benefits for the host community. An important point with respect to the nature of communities is made by Hall (2000) who notes that the emphasis on local, bottom-up approaches often associated with sustainable development appears to derive its legitimacy from an implicit assumption of the cohesion of local communities. However, communities are not homogeneous. They are made up of individuals and organizations who may well have different values, aims and objectives and who may or may not adhere, in varying degrees, to the dominant traditions of the community. These differing opinions can lead to conflict and power struggles. The political realities in the destination may well reflect that it is the local élite that controls tourism in a destination rather than a truly grass-roots community-based initiative.

How individuals and communities respond to tourism will play a part in how receptive the area is to tourists. The geographic setting and the strength of the local culture have roles to play. The arrival of a bus load of tourists in a remote village will have a potentially greater impact than the arrival of those same tourists in an urban area that is used to seeing tourists on a regular basis. At a broad level, Mathieson and Wall (2006) suggest that tourism may be viewed as the interaction of three types of cultures: the destination culture, the cultures of the visitors' origins and the tourist culture. The tourist culture reflects that although tourists may have different backgrounds, they often use the same facilities, visit similar sites and exhibit common behaviours. The three different cultures are not all homogeneous and they interact in a diversity of ways (Wall and Mathieson 2006). The diversity of the tourist culture will be explored further in Chapter 6.

Interactions between tourism and communities

Upon arrival in a community, tourists will encounter and interact with local residents; however, the nature of their trip and the type of accommodation and transportation they use will often determine the level of contact. The locals whom they encounter may be originally from the destination or may have migrated to the area in search of employment. Those who are employed in tourism or who benefit indirectly from tourism may be positive towards the industry while others may view tourism as an inconvenience or major problem to the community. Interaction may occur in the 'front stage' where a more formal service, performance or demonstration is given (see Plate 5.1), or the interaction may occur 'backstage' whereby

Plate 5.1 ***Argentina, Estancia Santa Susana, near Buenos Aires: Tour guide at a historic ranch presents traditional implements. Refreshments to be served to the tourists are on the side table.***

Source: Tom and Hazel Telfer

the tourist is allowed a view into the 'real life' of the residents, which may or may not be anything like the formal performance. Smith (1977 and subsequent editions) explored the nature of the complex interactions that occur between 'hosts' and 'guests' and the resulting impacts. These impacts are addressed in greater detail in Chapter 7. In a review of the literature, Burns and Holden (1995) summarized some of the main concerns relating to the interaction of the international tourism industry with developing countries. Table 5.1 presents those concerns with corresponding concerns for communities.

Although a number of concerns are outlined in Table 5.1, it is also important to note that tourism can also have positive impacts on communities such as reviving culture, creating a sense of empowerment and generating income. In the face of globalization there has also been a resurgence of regional or local identity as mentioned in Chapter 4. Citing the work of Ray (1998), Richards and Hall (2000a: 4) state: 'regions on the periphery of the global economy are asserting

Table 5.1 Concerns over tourism interaction in developing countries and implications for communities

Concerns over tourism	**Potential implications for communities**
Tourism development creates 'islands of affluence' in the midst of poverty	Local communities cut off from potential linkages to the tourism industry; possible resentment; migration may occur to communities near tourism sites for employment altering community structure; unequal sharing of benefits
Scarce national resources used for the enjoyment of wealthy foreign tourists	Loss of local resources taken for tourists such as water, land; communities may need to adjust to new resource availability
Impact of the demonstration effect on the local population	Some members of the community may adopt tourists' behaviour patterns, turning them away from traditional patterns of behaviour; local community traditions may be under threat
Economic multipliers, the main tool for measuring economic impacts, are controversial and unreliable	Economic benefits may not be as large as first expected; distribution of financial gains may not benefit entire community
Commercialization of culture and lifestyles	Community change to focus on high consumption lifestyle; culture and cultural artefacts are turned into commodities for sale; may revitalize culture
Benefits are probably going to foreign companies or local élite	Communities are controlled by local élites or foreign companies; few benefits to the local communities; challenges for communities to benefit from the industry
Control of international tourism is external to the destination and defined by transnational tourism corporations	Loss of control; limited opportunities for communities to participate in the tourism planning and development process; if controlled locally community-based tourism still has to interact with the international corporations who supply tourists

Source: left column, 'Concerns over tourism', after Burns and Holden (1995).

their identity as a means of preserving their cultural identity and developing their socio-economic potential.' The selling of regional or local identity has become part of the tourism product. Hall (2002) suggests there is an increasing recognition of the intellectual property dimension of tourism associated with regional characteristics.

Response to tourism by communities

Responses to tourism in a community can come from a variety of sources including individuals, community groups, business operators, non-governmental organizations (NGOs), environmental groups and the government, to name a few. Individuals may oppose tourism development while other community groups may band together to launch a community-based tourism project. In 1982, Mathieson and Wall published one of the first tourism texts entitled *Tourism: Economic, Physical and Social Impacts*. In 2006 the authors published an updated edition of the book and added the terms *Change* and *Opportunities* to the title in order to highlight the fact that not only are communities impacted by tourism development, but they also respond to the changes that tourism can bring. The term *consequence* is adopted in the new book to refer to the changes that occur as a result of tourism, since it has fewer negative connotations than the term *impacts*. Communities often seek to attract tourists and tourism developers and so development is often sought rather than imposed (Wall and Mathieson 2006).

In terms of the nature of communities, Wall and Mathieson (2006) stress that tourism takes many forms, and communities that embrace tourism have diverse characteristics. As a result, the consequences of tourism are highly contingent reflecting the specific forms and locations. The authors conclude, therefore, that making generalizations about the impacts in a community is very difficult, since both the type of tourism and characteristics of the community need to be taken into account before speculating on the consequences of tourism (Wall and Mathieson 2006). Murphy (1985) suggests that how a community responds to the opportunities and challenges of tourism depends, to a large degree, on its attitudes towards the industry. He notes that attitudes are personal and complex; however, in terms of community attitudes there are three main determinants. The first is the type of contact that exists between resident and visitor. The second is the relative importance of the industry to the individual and the community and the third is a tolerance threshold. This is the resident receptiveness expected in relation to the volume of business a specific destination can handle.

Models of attitudes towards tourism

In a detailed review of the literature, McGehee and Andereck (2004) suggest that residents' attitudes towards tourism is one of the most systematic and well-studied areas of tourism. Early studies adopted the perspective that communities were relatively homogeneous places and later research recognized that communities are heterogeneous with a wide variety of attitudes. Reaction to any tourism development will typically range across a continuum from acceptance to rejection and, according to attitude, certain behaviours may occur. Over time a number of models have been developed to articulate the responses of communities to tourism. Doxey (1976) suggests that attitudes towards tourism in communities go through a series of stages which include euphoria, apathy, irritation, antagonism, and a final stage when a community is undermined and what drew tourists initially no longer has the same attraction. The interaction between 'hosts' and 'guests' gradually becomes more formalized over time. This model, however, has received criticism on two fronts. The first is the inevitability of attitudes moving from positive to negative when the opposite may be true and, second, it may be misleading for communities to have dominant attitudes (Wall and Mathieson 2006). Butler (1975) adopted the attitude framework on cultural interaction to tourism from Bjorklund and Philbrick (1972), which indicates that the attitudes and behaviours of groups or individuals will be either positive or negative and active or passive (Wall and Mathieson 2006). As an example, entrepreneurs who are financially involved in tourism may promote the industry aggressively, while a small but highly vocal group uninvolved in tourism may lead aggressive opposition (Mathieson and Wall 2006). A key consideration here is to what extent opposition is permitted in differing countries. Dogan (1989) conducted a cross-cultural study of European tourists holidaying in Turkey and examined the response to tourism ranging from active resistance to the adoption of Western culture. Dogan (1989) proposed the following categories: resistance, retreatism, boundary maintenance and adoption. In a similar fashion, Ap and Crompton (1993) examined interactions of tourists and locals from the same culture and they devised the embracement–withdrawal continuum where responses fell into one of the following four strategies: embracement, tolerance, adjustment and withdrawal. These categories are explored further in Chapter 7 on the impacts of tourism (see Figure 7.6). The theoretical base for many studies has been social exchange theory, which suggests that people evaluate an exchange based on costs and benefits incurred as a result of the exchange (McGehee and Andereck 2004). Therefore, those residents who perceive themselves as benefiting from tourism will most likely view it positively, while residents who perceive themselves as incurring costs will be negative towards tourism. McGehee and Andereck (2004) do, however, note that there has been mixed support for social exchange theory in the studies of residents' attitudes.

Sustainable development and community involvement in tourism

In a review of the literature, Choi and Sirakaya (2006: 1275) state: 'sustainable development for community tourism should aim to improve the residents' quality of life by optimising local economic benefits, by protecting the natural and built environment and provide a high quality experience for visitors.' As indicated in Chapters 2 and 4, there have been a number of calls within tourism for approaches that incorporate the local community in the planning and development process (e.g. Jamal and Getz 1995). Murphy (1985) argues that the industry holds great potential for social and economic benefits if the planning is redirected from a pure business and development approach to one that is a more open and community-oriented approach that views tourism as a local resource. More recently, Jamal and

Jamrozy (2006) proposed an integrated destination management framework that brings in the destination's ecological-human communities as equitable and integrated members of goal-setting and planning-marketing. In the context of ecotourism, Jamal *et al.* (2006) call for a shift towards more of a social–cultural paradigm based on participatory democracy.

The importance of partnerships and the involvement of stakeholders have also been explored in the context of sustainable tourism (Jamal and Getz 1995; de Araujo and Bramwell 1999; Bramwell and Lane 2000). Partnerships are especially important in community-based tourism in developing countries where funds may be limited. Different organizations need to work together to effectively develop, promote and operate their product. De Araujo and Bramwell (1999) examined stakeholder assessment in tourism planning in ten municipalities in Alagos State in the north-east part of Brazil. The area is economically poor and is part of a larger tourism plan that is seeking to use tourism as a regional development tool. The authors argue that with inadequate involvement of the affected parties, the potential for conflict and reinforcement of inequalities can increase. They also comment that acceptance of a plan may be enhanced even by those affected by it if they have been involved in the planning process. The challenge with any stakeholder involvement, as will be illustrated later, is in managing a potentially time-consuming process.

Agenda 21 is the action plan adopted for sustainable development at the United Nations Conference on Environment and Development in 1992 held in Rio de Janeiro. This was a move, in part, to take the next step from the 1987 World Commission on Environment and Development, which produced *Our Common Future* and generated one of the most cited definitions of sustainable development. Chapter 28 of Agenda 21, which became known as Local Agenda 21, focuses on the role local political authorities can adopt in introducing a comprehensive planning process aimed at promoting sustainable development in their locality (Baker 2006). Local authorities were identified, as they have specific and significant environmental management functions in the following areas:

- Developing and maintaining local, economic, social and environmental infrastructure;
- Overseeing planning and regulations;
- Implementing national environmental policies and regulations;
- Establishing local environmental policies and regulations (Baker 2006: 106).

The framework of Local Agenda 21 establishes a possible path that promotes sustainability and encourages local involvement in the planning process. Government can act as a facilitator in the development of community-based tourism through the provision of start-up funds or training programmes. Subsequent international environmental conferences have re-emphasized the importance of Local Agenda 21. At the World Summit on Sustainable Development in Johannesburg in 2002, for example, a new toolbox of quantifiable actions was established to follow Local Agenda 21 titled Local Action 21 (Baker 2006). As extensive public involvement is an integral part of the Local Agenda 21 process, it helps establish a sense of purpose resulting in the community having new confidence in its ability to shape its future (Baker 2006). In resource management there are increased calls for co-operative management, which involves sharing rights between government and civil society (Plummer and Fitzgibbon 2004). Jackson and Morpeth (1999) argue that Local Agenda 21 may be used as a potential mechanism for implementing sustainable tourism. The World Tourism Organization along with the World Travel and Tourism Council (1996) also focused on the importance of local communities in their document *Agenda 21 for the Travel and Tourism Industry*. Jackson and Morpeth (1999: 33) suggest that 'sustainable tourism developments do offer genuine hope for marginalized communities, and we can find evidence of where those communities are being enabled to play a meaningful role in delivering tourism initiatives which are appropriate to those specific localities and regions'.

The goals of initiatives such as Local Agenda 21 and its possible adoption in tourism are, in part, to promote community development. The United Nations tentatively defined community development as 'a process designated to create conditions of economic and social progress for the whole community with its active participation and the fullest possible reliance on the community's initiative' (United Nations 1955: 6). Nozick (1993) developed a set of principles of sustainable community development which include:

- Economic self-reliance;
- Ecological sustainability;
- Community control;
- Meeting individual needs;
- Building a community culture.

Richards and Hall (2000a) effectively make the connection of the sustainability efforts of local communities with the global: 'Local communities become not only important in terms of actions taken to preserve their own immediate environment, but also form part of wider alliances to preserve the environment globally (act local, think global)' (Richards and Hall 2000a). The authors also highlight the role of NGOs and other pressure groups having like-minded members who are environmentally aware, as they themselves can be viewed as communities.

While the potential is there, caution is raised by Jackson and Morpeth (1999) in the context of tourism, and Baker (2006) in the broader context of sustainable development, about the difficulty in implementing Local Agenda 21 plans. Surveys of Local Agenda 21 initiatives indicate that there are varying levels of stakeholder involvement (Baker 2006). In developing countries in particular, there is often a lack of highly developed civil society structures, which acts as a barrier to some elements of Local Agenda 21 initiatives (Baker 2006). Even though there are continued calls for local involvement, Lui and Wall (2006) argue for the importance of moving beyond rhetoric and understanding the capabilities of locals to actually participate. They call for an increased recognition of human resource planning: 'If tourism is really to be a "passport to development" and a means to enhance the lives of destination residents then greater attention must be given in tourism plans to their needs and capabilities' (Lui and Wall 2006: 169). Milne and Ewing (2004) explore community participation in tourism and suggest that in the Caribbean and elsewhere, regardless of political systems, evolving stakeholder relationships and access to information technology, participation will not occur in a sustainable fashion unless people have the will and interest to take control of their own destinies. Later in the chapter the barriers in terms of participation in the planning process are addressed more directly in the context of tourism.

Community-based tourism

Community-based tourism is one type of tourism that incorporates high levels of community involvement under the sustainability umbrella. It is often viewed at the opposite end of the spectrum from large-scale, all-inclusive, mass tourism resorts owned by corporations that have limited economic linkages to communities with, perhaps, some residents of the local community being hired in low-skilled and low-payed jobs (Hatton 1999). Community-based tourism is local tourism developed in local communities in innovative ways by various individuals and groups, small business owners, entrepreneurs, local associations and governments (Hatton 1999). Funding can come from a variety of sources including international donors. The Asian Development Bank, for example, is providing funds through its Mekong Tourism Development Programme to community-based and pro-poor initiatives and, through this programme, funds have been provided to provinces in Lao PDR (Harrison and Schipani 2007). Community-based tourism can also be linked to some forms of indigenous tourism. Indigenous tourism is defined as 'tourism activity in which indigenous people are directly involved either through control/and or having their culture serve as the essence of the attraction' (Hinch and Butler 1996).

There are several goals of community-based tourism (Hatton 1999). The first is that it is socially sustainable. Tourism activities are developed and operated for the most part by local community members and participation is encouraged. In addition, the revenues are directed towards the community through various potential ways such as co-ops, joint-venture community associations, businesses that employ local people, or to a range of entrepreneurs starting up or operating small and medium-sized enterprises. The second major goal for community-based tourism is respect for local culture, heritage and traditions. It has been suggested that community-based tourism can also reinforce or rescue local culture, heritage and traditions. In addition, respect is implied for natural heritage especially when the environment is part of the attraction. To the north of the city of Yogyakarta, Indonesia, is the village of Bangunkerto, the site of a community-based agri-tourism initiative (see Plate 5.2). With the help of local governments, the community switched their crops to higher quality salak fruit and opened up an agritourism site for tourists. Based on interviews, it was found that all aspects of the site are controlled at the local level and the operation has strengthened local identity (Telfer 2000).

Based on a brief survey of community-based tourism in members of the Asian-Pacific Economic Cooperation (APEC), Hatton (1999) identified a number of recurring themes. The first theme is focused on why community-based tourism

Plate 5.2 ***Indonesia, village of Bangunkerto: The site of a community-based agritourism project based on tours through a salak plantation. These women are laying stones by hand on the road bed in order to prepare it for paving. This is being done so that tour buses are better able to get through the community.***

started in the various destinations. The common factor identified is the expectation of economic gain, which in some cases is directly related to need. The second theme is leadership linked to the initiative from one person, a small group or, in some cases, the government. Cultural heritage is often one of the most important aspects of community-based tourism and in many cases is the attraction for the tourists, and so it was identified as a third theme. The natural environment is also a key theme for many communities where tourists are drawn to experience the environment. See Box 5.1 for a discussion on community-based ecotourism in Cuba (see Plate 5.3 and Plate 5.4). The fifth theme is that community-based tourism is linked to the growth of employment opportunities, particularly for women, young people and indigenous peoples. Finally, in community-based tourism there is an emerging theme where corporations and communities are starting to work together.

In the context of indigenous tourism, Hinch and Butler (1996) discuss the key aspect of control of the industry. 'Whoever has control can generally determine such critical factors as the scale, speed and nature of development' (Hinch and Butler 1996). The authors have established a framework for indigenous tourism that looks at the level of control the indigenous people have over tourism enterprises and also whether there is an indigenous theme featured in the attraction. Attractions, services and infrastructure controlled by indigenous people and developed around indigenous themes represent the strongest manifestation of

Box 5.1

Community-based ecotourism in Cuba

Traditionally, local participation in the tourism planning and development process in Cuba has been limited due to government controls. With the collapse of the Soviet bloc, Cuba turned to tourism to generate foreign exchange. The government allowed foreign capital to be invested in joint ventures. With a lack of a participatory process, the government has been able to 'fast-track' beach resort development in places such as Cayo Coco and Cayo Guillermo as well as other locations. However, the government has been attempting to diversify the product away from mass tourism towards more niche products such as health, nature and cultural tourism. This brings tourists into more direct contact with locals and the pressure for increased levels of participation is growing (Milne and Ewing 2004). One example of nascent local participation is La Moka Ecolodge (26 rooms) built in 1994 next to the community of Las Terrazas in Sierra del Rosaria, a biosphere reserve. The community-managed hotel was the idea of the Minister of Tourism; however, the local community was consulted about the project. Promoted as a nature and ecotourism destination, the Tourist Map for the Reserve states the following:

> Las Terrazas Tourist Complex is a rural experience of sustainable development that combines 5000 ha of forest in the heart of Sierra del Rosario Reserve of the Biosphere and a working community of 890 inhabitants, offers a unique product in which the endemic Flora and Fauna, hotel services and social experiences will give you memories to be treasured.
>
> (Hecho En GeoCuba, 1997)

The hotel is small scale and was designed to fit into the surrounding environment. In places, trees were left intact such as the one which grows up through the main lobby of the hotel and out through the roof. Excursions into the biosphere include guided nature walks, bird-watching, trekking, hiking, horseback riding and mountain bike tours as well as visits to the partially restored remains of a coffee plantation. Residents of the community have the opportunity to work in the hotel, or to sell crafts to the tourists, while others work in the park (Telfer 2001). New community facilities were built close to the hotel and a contain a small range of services for the residents such as shops and a health clinic. The community is scheduled to repay the government investment ($6 million) over a 15- to 20-year period and 40 per cent of the profits go into a community development fund overseen by the neighbourhood Committee for the Defence of the Revolution, with 10 per cent going to the community health clinic (Milne and Ewing 2004).

Sources: Telfer (2002); Milne and Ewing (2004).

indigenous tourism. Smith (1996) identifies four interrelated elements in indigenous tourism:

- The geographic setting (habitat);
- The ethnographic traditions (heritage);
- The effects of acculturation (history);
- The marketable handicrafts.

As the culture itself becomes the attraction, there are concerns raised over authenticity when the culture is opened up to tourists. Indigenous culture can be sold to tourists through cultural performances and souvenirs, which may change over time

Plate 5.3 ***Cuba, La Moka Ecolodge: Visitors approach the main entrance to the hotel. Note the architectural design of the roof.***

Plate 5.4 ***Cuba, Las Terrazas: Located next to La Moka Ecolodge, the community of Las Terrazas is home to some of the people who work in the resort.***

to suit the tourists' needs. Yamamura (2005) examined the culture of the *Naxi* people in the old town of Lijiang, China, where new tourism-related merchandise is being produced in accordance with the traditional handicraft techniques. Smith (1996) notes that although there are conflicts among some indigenous people over the desirability of tourism in their home communities along with conflicts over the previously mentioned changes to handicrafts, there is an economic element to be considered as it can provide jobs and income. In addition, rural and indigenous peoples' environmental ethics may differ from more urban counterparts and they may view environmental controls and protection as contradictory, limiting their development to satisfy the desires of urban sophisticates (Butler 1993).

While community-based tourism has many positive points, these relatively small-scale tourism operations do face a number of challenges. Cleverdon and Kalisch (2000) suggest that one of the major challenges facing community-based tourism is the competition and threat posed by large-scale resorts in the vicinity. With their resources and marketing techniques, the larger operations have the potential to take business away from the small operators. The key challenge for the tourism industry and policy makers is to find a way for the large and small firms to co-exist and work together as part of an integrated local economic development policy (Cleverdon and Kalisch 2000). Other challenges relate to the long-term viability of a community-based project and whether it becomes controlled by just a few. In an analysis of power, Coles and Church (2007: 7) suggest that while 'equitable, fair and locally empowering forms of tourism production, governance and consumption remain the aspiration, inevitably they require interaction among human beings; in other words they are political processes and they are the subject of power relations among constituencies'. It is also important to remember that while sustainable community-based tourism may put some of the control of tourism in the hands of the local people, they must still interact with the external agents of the tourism industry, which is largely responsible for bringing in the tourists in the first place. Finally, Blackstock (2005) states that, from a community development perspective, community-based tourism has three major failings:

- It seeks to ensure the long-term survival of the tourism industry rather than social justice;
- It tends to treat the host community as a homogeneous bloc;
- It neglects the structural constraints to local control of the tourism industry.

Participation

If there is to be participation by communities in the planning process it is important to consider the actual level of participation. In an analysis of Local Agenda 21 initiatives in developing countries, Baker (2006) highlighted the structural difficulties for participation. There may also be difficulties for those in power to release control and allow the locals into the process. What then are the different levels or forms of participation? Arnstein (1969) developed a typology or ladder of citizen participation with eight rungs. The bottom two rungs are manipulation and therapy. These levels are described as non-participation and are more about allowing power holders to educate or cure participants. The third and fourth rungs of the ladder progress to tokenism allowing the have-nots to have a say. These rungs are informing and consultation, and citizens lack the power to ensure that those in power will heed their views. The fifth rung, placation, is a higher level of tokenism, as citizens can advise but the power holders still make the decisions. The last three rungs are levels of citizen power and increasing levels of decision-making clout. The sixth level is partnership, which allows for negotiation and trade-offs with those in power. The seventh rung is delegated power and the last rung is citizen control where the have-nots have the majority of decision-making seats or managerial control (Arnstein 1969).

In addressing the limitations of the typology, Arnstein (1969) acknowledges that it does not address the road-blocks to achieving genuine levels of participation. On the power holder side they include racism, paternalism, and resistance to the redistribution of power. On the have-nots side they include inadequacies of the poor community's socio-economic political infrastructure and knowledge base. In addition, there are the difficulties of organizing a representative and accountable citizens' group in the face of futility, alienation and distrust (Arnstein 1969).

Limits to participation in tourism

While, in theory, communities should be involved in the tourism planning process in developing countries, there can be significant barriers. Tosun (2000) indicates that the concept of the participatory development approach originated in the developed world. Concepts, theories or models created in the developed world will face challenges in their application to developing countries. Tosun (2000) identified operational, structural and cultural limits to community participation in the tourism development process in developing countries. These limits are often a reflection of the prevailing socio-political, economic and cultural structures in many developing countries. At the operational level, obstacles included the centralization of public administration of tourism development, lack of co-ordination between involved parties, and a lack of information available to locals in the tourist destination. Structural limitations involve institutional barriers, power structures, and legislative and economic systems. More specific items identified within the structural limitations include the problem of persuading professionals to adopt participatory tourism development, lack of expertise, élite domination, lack of an appropriate legal system, lack of trained human resources, a relatively high cost of community participation, and lack of financial resources (Tosun 2000). Finally, Tosun (2000) identifies cultural limitations such as as the restricted capacity of poor people, and apathy and low levels of awareness in the local communities. These constraints are similar to what Timothy (1999) found in Yogyakarta, Indonesia, and he grouped them under the following categories:

- Cultural and political traditions;
- Poor economic conditions;
- Lack of expertise;
- Lack of understanding by residents.

It is also important to consider the existing political structure of the destination to see if a more open process of community participation is allowed or possible. In the case of both Cuba (Milne and Ewing 2004) and China (Li 2004), significant barriers to a democratic of community-based tourism planning are apparent. In Yogyakarta, Indonesia, Timothy (1999) found that one of the most apparent traditions in Javanese culture is that of authority and reverence towards the people in positions of power or high social standing. This concept permeates from high-level political jurisdictions down to the village level and to the level of the family. In the case of the village head, for example, villagers view that person as authority and few would bypass the village head for receiving and giving advice, as this behaviour would be offensive and cause the village head to lose face.

Empowerment

The concept of empowerment is difficult to define; yet it is embraced by a wide range of organizations with diverse social aims, as it is attractive and viewed as politically correct (Scheyvens 2003). Arai (1996) outlined the following five concepts relating to empowerment based on the literature:

- It involves a change in capacity or control, or an increase in power and the ability to use power;
- It has multiple dimensions including psychological, economic, social and political change;
- It is a multilevel construct with changes occurring within an individual, group or community;
- It should be understood from a holistic perspective combining the above listed dimensions and levels;
- It is a process or framework that describes changes as an individual, group or community mobilizes towards increased citizen power.

Scheyvens (2003) used the same dimensions (psychological, economic, social and political) of empowerment as Arai (1996) in the context of tourism, which are explained below. If a community is empowered economically it will need to ensure access to productive resources in the tourism area. Scheyvens stresses that this is particularly important in the context of common property resources. If a park is established the local community needs to be able to benefit from the park, but if indigenous rights to harvest from the park are undermined then they will not be empowered economically. Social empowerment occurs when a community's sense of cohesion and integrity is confirmed or strengthened by being involved with tourism. Social empowerment is increased if money generated from tourism is used for social development projects such as improving water supply subsystems or health clinics. Social disempowerment occurs if there are some of the negative social impacts sometimes associated with tourism such as crime, displacement from traditional lands or prostitution. Psychological empowerment is a reflection of a community's confidence in its ability to participate equitably and effectively in tourism planning, development and management. Tourism that is sensitive to cultural norms and respects traditions can be empowering; however, tourism that undermines local culture will have a negative effect. Scheyvens (2003: 235–236) suggests that 'feelings of apathy, depression, disillusionment or confusion in the face of tourism development could suggest that psychological disempowerment has occurred'. Finally, Scheyvens (2003) explores political empowerment. If community members are politically empowered their voices are heard and should guide the tourism development process. The importance of diverse interest groups in a community is raised and all voices need to be heard through democratic processes or along more traditional lines of communication.

The power of communities in the face of tourism

An externally controlled tourism industry can be a powerful force for developing countries to deal with. The forces of globalization have multinationals searching for the lowest production costs, resulting in governments in developing countries offering various incentives to attract international developers. In developing countries, much of the planning is top-down with governments controlling the framework for the industry. Butler's (1980) model of resort development illustrates the various stages a resort could go through over time, following the pattern of a product life cycle in rising and falling and perhaps rising again. In the early stages of the model where there is limited tourism, it is suggested that communities have more control over development. As the destination evolves and external companies enter the destination, control is passed from the local community to external agents. Various levels of government may also intervene to try to use tourism in the area as part of a regional or national development plan which may mean relocation of communities, as illustrated below.

As tourism grows, changes and opportunities occur. Tourism also has the potential to make a positive impact in terms of promoting a sense of place and identity for a local community. Richards and Hall (2000a) state that communities are not simply victims of the globalization process and commodification, but they can also become centres of resistance as is discussed below. James (2006) suggests that one of the dominant trends in the current period is the tension between globalism and localism, and interest in local differences is also what attracts tourists to locations. The promotion and marketing of local or regional identity in terms of tourism can generate employment opportunities, as is illustrated later.

Relocation

In some cases, entire local communities have been displaced to make way for new tourism developments, with some receiving compensation while others do not. Long (1993) explored the complex challenges and the differing reactions in the seaside community of Santa Cruz Huatulco in Mexico (population 735 with 250 households) to the relocation of their community one kilometre inland. This was done to make way for the Las Bahais de Huatulco, a master-planned mega-resort development project by FONATUR. FONATUR used a number of social impact mitigation techniques; however, Long (1993) found that the original residents saw community changes in every aspect of their community. These included an influx of outsiders, a shift from an ocean environment to cement houses on dusty streets, and the development of new social classes, new occupations groups, and new organizations. As with any controversial development project, Long (1993) also commented that there were those who supported the project. This finding reinforces the fact that communities are not homogeneous and it is not correct to portray communities as simply the victims of globalization. In a more recent study, Wang and Wall (2005) explored the consequences of tourism-caused displacement of a minority community in Hainan, China, where tourism is being used as a regional development strategy and is being planned in a top-down manner. During the preparation period for the resettlement the people had expressed great stress about the displacement and rejected additional risk-taking activities. The actual relocation distance was short and, along with the integrated resettlement pattern, resulted in no obvious external settling change in this relocation. Following the move to the new village, stress and conservatism decreased, resulting in more innovative behaviours by the residents. Wang and Wall (2005) argue that if those being resettled are to get their share of benefits to tourism, improvements in planning are required as well as training opportunities and greater access to jobs.

Resistance to tourism

Kousis (2000) explored resistance movements to tourism and examined local environmental mobilizations against tourism developments in Greece, Spain and Portugal, while Routledge (2001) examined local resistance to tourism in Goa, India. In Goa, the expansion of luxury and charter tourism has increasingly involved investments from transnational corporations. The emphasis has shifted away from backpackers and charter tourism to luxury tourism, as it brings higher spending tourists. Routledge (2001) states that the development of Goa as a tourist site has had serious consequences for the state's people, ecology and political economy, and has transformed the coastal areas into dispensable space. Economic growth is considered of prime importance and environmental problems are secondary to growth (Roultedge 2001). The situation led to the emergence of a number of Goan opposition movements including citizens, activist village groups, and environmentalist non-governmental organizations concerned over tourism development. Routledge (2001) examined two organizations including the Goa Foundation and *Jagrut Goenkaranchi Fouz* (JGF) or Vigilant Goan's Army, and identified a number of resistance examples which ranged from information dissemination to village mobilizations and protests, demonstrations and road-blocks. Finally, Boissevan (1996) highlights that local residents have devised a range of strategies to prevent tourists from penetrating their back regions, which include:

- Covert resistance – low-key resistance of sulking, grumbling, obstruction, gossip;
- Hiding – withdrawal from tourists to celebrate;
- Fencing – fencing off private areas/rituals;

- Ritual – resurgence in celebrations helps protect against impacts of tourism and helps cope with change;
- Organized protest;
- Aggression.

Tourism and the informal sector

Part of the interesting dilemma of tourism development that has been highlighted in previous chapters is the debate over resort enclaves targeting mass tourists. While these types of development have been criticized for being isolated, in some cases it is the existence of these resort complexes that enable small-scale community-based projects to open up, taking advantage of the tourists already in the destination. The informal sector plays a major role in developing countries. In the context of urban areas, Pal (1994: 79–80) suggests the potential for

> small-scale informal sectors to provide employment and generate income for countless urban dwellers in burgeoning cities of the developing world is considerable. Indeed, given the surplus of labour and high levels of poverty in urban areas, the encouragement of these sectors should be a cornerstone of any policy promoting sustainable development.

For tourism in rural or urban areas, the response of the informal sector will be important for the overall economic impact of the industry. In studying tourism employment in Bali, Cukier (2002) found evidence of migration to the tourism area of Bali from other islands and those in the informal sector saw their current employment as the means to acquire the skills to move to the formal sector. Citing studies in Yogakarta, Indonesia, and elsewhere in South Asia, Shah (2000) comments that domestic or regional tourists tend to buy more from local vendors than do Western tourists.

How communities and individuals respond to tourism, whether through resistance or trying to take advantage of the possible employment opportunities, will, in part, dictate the role tourism can play in development. The chapter now shifts to examine several current issues in tourism as they relate to communities including the role of NGOs, fair trade and tourism, pro-poor tourism, volunteer tourism, and gender and community development. As mentioned in the introduction, some of these concepts, programmes and agents may promote development through the use of tourism; yet they all face significant challenges.

The role of NGOs in community tourism

Before examining the role of NGOs in community-based tourism it is important to define civil society of which NGOs can be identified as being a part. Ottaway (2005) notes that defining civil society is complex due to the nature of the organizations involved. Drawing on the often-cited definition by Hegel (1821), Ottaway defines civil society as comprising the entire realm of voluntary associations between the family and the state. These voluntary organizations have many different forms, ranging from small, informal groups to large, bureaucratic organizations with large budgets. In developing countries these organizations may have a very narrow focus such as providing aid to AIDS orphans, or at the other extreme these organizations are linked to transnational networks with goals such as trying to influence World Bank operations (Ottaway 2005). These groups may try to provide services the state is not able to deliver or help citizens resist pressures from predatory governments (Ottaway 2005). In Agenda 21 there are calls for both the United Nations and governments to involve NGOs in making policy and decisions on sustainable development.

Non-governmental organizations (NGOs) are playing an increasing role in influencing tourism. Baker (2006: 9) defines NGOs as 'organisations operating at the national and increasingly, at the international level, which have administrative structures, budgets and formal members and which are non-profit-making'. However, they can also operate at a very local scale. NGOs have taken on a number of different roles within developing countries including providing development relief, raising awareness over specific issues such as environmental concerns or sex tourism, lobbying governments, assisting local community with projects, and assisting building community capacity to name just a few. Some NGOs also help to coordinate volunteer tourism, which will be discussed below. NGOs are also involved in the movement towards 'new practices of environmental governance' (Baker 2006). Environmental concerns such as climate change, biodiversity loss and deforestation have resulted in issues that cross political borders, raising questions about traditional environmental governance practices. The new practices have NGOs alongside states and international organizations, as well as the use of a wide range of policy instruments (legal, voluntary and market instruments) and normative and governance principles to promote sustainable development (Baker 2006).

Tourism Concern, based in the UK, is well known for its campaigns to raise awareness over a variety of issues with respect to tourism. A few of their projects include raising awareness over the poor treatment of employees in the tourism sector in developing countries, issues of human rights abuses and tourism, promoting ethical and responsible tourism and campaigning against specific tourism developments for environmental concerns, to name a few. Tourism Concern has called for a boycott on travel to Burma due in part to human rights violations in the country. The Ecumenical Coalition on Third World Tourism, based in Hong Kong, is also conducting a number of campaigns against the abuses of the tourism industry.

In the context of communities, Burns (1999b) suggests that NGOs can often act as a bridge to promote cooperation within communities, establishing initial links with the local and regional government tourism sector to form partnerships. While some NGOs work across international borders, many are very small and work in specific countries or communities. Barkin and Bouchez (2002) document the work of the Centre for Ecological Support (CSE) on the Pacific coast of the Mexican state of Oaxaca. The area is the location of a mega resort developed by FONATUR, the Mexican Tourism Development Fund. Approximately 30 km of the coast was expropriated by FONATUR. The authors state that the tourism development and accompanying infrastructure integrated the previously isolated community into the international market, sparking a self-reinforcing cycle of speculation and investment. This accelerated process of social and spatial polarization impoverished the native populations and created tensions. The community was also hit by hurricane Paulina in October 1997. The CSE was originally created in 1993, after the first large hotels were inaugurated in the new resort, to promote regional development. Work had begun on a resource management plan for sustainable development, and the NGO worked with native communities to regenerate smaller river basins to promote community welfare, including replanting trees with commercial and cultural value. The reforestation was complemented by a variety of initiatives, including ecotourism. By creating a favourable environment they would possibly attract tourists. The ecotourism project was designed to be owned and managed by indigenous communities participating in the programme and which were sensitive to the natural heritage they were rescuing and protecting.

One of the main questions addressed by this book is whether tourism can contribute to development broadly defined. Archabald and Naughton-Treves (2001) examined tourism revenue sharing around national parks in Western Uganda with neighbouring local communities. In the case of Bwindi Impenetrable National Park, two international NGOs (International Gorilla Conservation Programme (IGCP) and Cooperative for Assistance and Relief Everywhere (CARE)) worked with the Uganda Wildlife Authority in the initial stages of the tourism revenue programme on issues such as strategic planning assistance and workshops, as well as technical, logistical and monetary support for community training. Mountain gorilla ecotourism was part of the park's offerings. The tourism revenue-sharing programme funded community development projects in 19 of the 21 parishes bordering the park and each parish received approximately US $4,000, with the money going to building primary schools, health clinics or roads. While the authors note the programme has been suspended, it does highlight the possibilities of tourism revenue-sharing programmes for some of the broader aspects of development.

There are a variety of other international NGOs such as the WWF, IUCN and Conservation International that have been involved in a number of environmental and conservation projects including tourism and ecotourism. On a cautionary note, however, Mowforth and Munt (1998) raise the concerns that through coordinating links to the World Bank some international NGOs have adopted an approach advocating corporate schemes for environmental projects giving 'total management control' to the private or NGO sector. This leads to the possibility that they may not be fully aware of the details of local village conservation movements. The rise of NGOs also illustrates the changing nature of governance, as states are not only dealing with multinational corporations; they are also dealing with, in some cases, international NGOs.

Fair trade in tourism

Local communities face challenges trying to interact with the highly competitive and international nature of tourism. One concept that is gaining more attention is fair trade. The practice of fair trade has often been associated with the production of foodstuffs (e.g. coffee) and handicrafts, and is often linked to aid programmes in developing countries (Evans and Cleverdon 2000). The idea is that local producers in developing countries receive not only support but also a fair price for their products. If this concept can be translated into tourism, local communities will benefit in a number of ways. Evans and Cleverdon (2000), citing the work of Barratt Brown (1993), outline the main concepts of fair trade as:

- Improving working conditions;
- Improving production and marketing of goods;
- Implementing premium pricing, training and investment;
- Minimizing economic leakage;
- Widening the distribution of economic benefits;
- Guaranteeing price stability;
- Guaranteeing more sustained income.

Fair trade is distinguished from free trade as it fights against poverty in developing countries and brings the consumer into contact with the producer. This is especially so for tourism that occurs in local communities in developing countries. It is also different from free trade as it targets small-scale producers, and the trading method is designed to strengthen the partner's position in developing countries. This includes assistance from developed countries in product development, control over image and representation, and access to marketing techniques (Cleverdon and Kalisch 2000). It is important to keep in mind that the tourism industry consists of predominantly small and medium-sized enterprises (SMEs) (Richards and Hall 2000b). These may range from a family opening their home to tourists in village tourism or an individual waiting at a historical monument for tourists to arrive to act as their guide. The tourism industry relies on innovation and so the entrepreneurs are an important part of the industry and can represent a crucial alternative to transnational companies (Richards and Hall 2000b). If these small firms receive the main benefits from the fair trade process, which include 'transparency of trading operations, commitment to long-term relationships and the payment of prices that reflect an equitable return for the input provided' (Cleverdon 2001: 348), they will be better able to compete in the tourism market.

Although there is a great deal of opportunity in this area, there are a number of differences between trade in primary products such as coffee and tourism that will present challenges to implementing fair trade in tourism. Some of these differences are as follows:

- Fair trade organizations are non-profit while tourism organizations operate in a highly competitive market seeking profits.
- The tourism product is intangible and invisible, and involves multi-sector activity.
- The success of tourism depends on low, flexible prices while primary products, such as coffee, have a 'world price'.
- Developing countries are used to exporting primary products; however, in many grass-roots communities there is little experience of using tourism as an export item.

- Tourism is typically based on competitive entrepreneurship and the collective organization of small-scale tourism providers is a new concept.
- Tourism is consumed in the destination, bringing social and cultural intrusion (Cleverdon and Kalish 2000).

While these differences may provide challenges for larger tour operators to meet, there may be opportunities for smaller operators to create a strategic advantage by incorporating fair trade principles into their business operations (Chandler 1999 in Cleverdon 2001).

Pro-poor tourism

Since the 1992 Earth Summit, sustainability efforts had been heavily biased towards environmental considerations with tourism being seen primarily as a form of sustainable use of natural resources and a way to enhance conservation (Rogerson 2006). Rogerson (2006) argues that what was missing or downplayed was concerted attention to the impacts of tourism upon the poor and, until recently, few governments in developing countries considered linking tourism development directly to poverty reduction. It is not surprising, then, with poverty alleviation strongly evident in programmes such as the United Nations Millennium Development Goals, that tourism should be viewed as a strategy to alleviate poverty. Often referred to as pro-poor tourism, it 'focuses on how tourism affects the livelihoods of the poor and how positive impacts can be enhanced through sets of interventions or strategies for pro-poor tourism' (Rogerson 2006). Pro-poor should not be regarded as a specific type of tourism but rather as an overall approach to poverty reduction through tourism. The concept has similarities to fair trade in tourism in its focus on poverty. Pro-poor tourism strategies have the potential to be implemented in a variety of types of tourism whether they are large scale or small scale, urban or rural. More specific types of tourism, such as community-based tourism or ecotourism, have an emphasis on enhancing local participation in the benefits of tourism and so have affinities with pro-poor tourism, but would also need to have strategies to assist the poor. The concept of pro-poor tourism has received increased attention with the UNWTO launch of the ST–EP programme (Sustainable Tourism – Eliminating Poverty) at the World Summit on Sustainable Development in Johannesburg in 2002. The journal *Current Issues in Tourism* has dedicated a special issue to the topic (see vol. 10, nos 2 and 3, 2007).

There are a number of advantages inherent in tourism that make it attractive for promoting pro-poor growth as identified by Rogerson (2006). There is wide scope for participation in tourism, including the informal sector. The customers come to the product, creating opportunities for linkages to other sectors. Tourism is highly dependent upon natural capital (wildlife and scenery), and culture and the poor have some of these assets. Tourism is able to be more labour intensive than manufacturing. Finally, compared to many other economic sectors, a greater proportion of benefits from tourism in jobs and entrepreneurship opportunities accrue to women.

It was pointed out at the opening of this chapter that communities are not homogeneous. Ashley *et al.* (2000) also stress the fact that the poor are not homogeneous and the impacts of tourism (positive and negative) will inevitably be distributed unevenly among poor groups reflecting differing patterns of assets, activities opportunities and choices. Ashley *et al.* (2000) also comment that it is often community tourism that is thought of as the main avenue for the poor to participate through community-run lodges, campsites or craft centres often supported by NGOs, yet poor individuals participate in all types of tourism through self-employment such as hawking and casual labour. The authors indicate that although more research is needed on the participation of the poor in various tourism market segments, they stress the importance of the domestic/regional tourism market. With the focus on livelihoods, the question turns to what policies or programmes would contribute to pro-poor tourism. Ashley *et al.* (2000) outline the following strategies to enhance economic participation of the poor in tourism:

- Education and training targeted at the poor (particularly women) to take up employment and self-employment.
- Expand access to micro-finance.
- Recognize and support organizations of poor producers.
- Develop core tourism assets and infrastructures in relatively poor areas where commercially viable product exists.
- Strengthen local tenure rights over land, wildlife, cultural heritage, access to scenic destinations and other tourism resources.
- Use planning to encourage investors to develop strategies to assist the poor.
- Minimize red tape that excludes the least skilled.
- Enhance vendors' access to tourists.
- Business support to improve quality, reliability of supply, transport links.
- Incorporate domestic/regional tourism and independent tourism into the planning process.
- Avoid excessive focus on international all-inclusives.
- Recognize the importance of the informal sector in the planning process.

One of the key items above is the expansion of micro-finance to help the poor who do not have the capital to launch new initiatives (see Vargas 2000). The importance of this was demonstrated in 2006 when the Noble Peace Prize was awarded to Muhammad Yunus and Grameen Bank in Bangladesh for work on micro-credits as an instrument against poverty. It is interesting that the Peace Prize went to an economic initiative. An important challenge associated with micro-credit schemes, especially those funded from international agencies, is that in some communities in developing countries the nature of development may be defined differently than that of the West, and so macroeconomic indices used by international funding agencies to assess the micro-credit schemes may not work (Horan 2002). Horan (2002) found that a group of women in Tonga, who received funding through a micro-credit scheme for textiles that were to be sold to tourists, were engaged in their notion of development known as *fakalakalaka*, which is a broader form of development than can be measured by economic indicators. It involves all areas of the individual (physical, mental and spiritual) and extends to personal relationships, family dynamics and the development of the community. The women used the money to make ceremonial textiles for their local culture rather than textiles for tourists. Although there was a low default rate on the loans the international lenders did not recognize that development was occurring. The projects funded by micro-credits need to fit with local conditions.

In the democratic era of South Africa, tourism has become an essential sector for national reconstruction. The government has developed a wide range of interventions to involve poor communities in tourism. Through a variety of different funds and programmes, the government has encouraged small-scale entrepreneurs to develop new products as well as the tourism industry to reach out to poor local communities and incorporate their workers, products and services into the tourism industry (Rogerson 2006). See Box 5.2 for a more detailed discussion of pro-poor tourism initiatives in South Africa (see also Plate 5.5).

Box 5.2

Pro-poor tourism in South Africa

In the democratic era in South Africa, tourism is seen as a development tool. Rogerson (2006) has identified a number of government programmes and pro-poor tourism pilot projects. Beginning with the release of a White Paper on the Development and Promotion of Tourism in 1996, there has been a focus on responsible and sustainable tourism and an important recognition of the need for communities to be involved and to benefit from tourism. There is 'a strong emphasis in tourism planning upon job creation and enterprise development in support of the country's neglected black communities' (Rogerson 2006). The government has instituted a number of different programmes to support the involvement of poor communities in tourism. These initiatives include funding for infrastructure and new product development, such as cultural tourism and handicrafts in rural areas and township tourism in urban areas. Funds have also been set up for assisting tourism entrepreneurs and training programmes. A Fair Trade in Tourism South Africa brand and trademark have also been established. Guides for responsible tourism and guides of good practice have been made available to operators. Some of the recommendations in these guides include prioritizing opportunities for local communities, tourism businesses buying locally made goods and using locally available services, and stipulating that local people should be hired. Some ecotourism operators, for example, have been expanding benefits to local communities. Rogerson (2006) identifies township tourism in Alexandra Township on the edge of Johannesburg as part of the pro-poor initiative. As a result of the apartheid government the Township is one of the most impoverished and densely populated in South Africa. It has been the centre of Township tourism, bringing tourists to sites of significance to the anti-apartheid movement and generating an understanding of issues of poverty and historic oppression. As well as being part of a campaign against urban poverty, Alexandra Township tourism has been part of a pro-poor pilot project linking it to the Southern Sun Hotels Group. A series of existing programmes and proposals to link the hotel group to the community were developed/proposed, including outsourcing from the Township, marketing tours to the Township through the hotels, recycling guest amenities to Township entrepreneurs to make into crafts, and the sale of locally produced goods (souvenirs) to hotel guests. Rogerson (2006) reports that difficulties were encountered in the project, highlighting the challenges in maintaining partnerships.

Source: Rogerson (2006).

Plate 5.5 ***South Africa, Township near Pretoria: Tourists visit a Township.***

While pro-poor tourism represents a step towards increased involvement of the local community, Mowforth and Munt (1998: 272) argue that given the small size, it is not 'a tool for eliminating nor necessarily alleviating absolute poverty, but rather is principally a measure for making some sections of poorer communities "better-off" and of reducing vulnerability of poorer groups to shocks (such as hunger)'. It has also been pointed out that it can place considerable demands on tourism stakeholders who operate in a market environment to adopt strategies to maximize benefits for the local poor, which may include higher wages that need to be paid regardless of business cycles (Chock *et al.* 2007). These authors go on to point out that successful pro-poor tourism relies, to a large extent, on the altruism of non-poor tourism stakeholders to move the industry towards increasing benefits along with reducing the costs for the poor. Hall (2007b) summarizes the critics of pro-poor tourism as suggesting it is 'another form of neo-liberalism that fails to address the structural reasons for the north–south divide, as well as internal divides within developing countries'. Schilcher (2007) examines the ideological aspects of pro-poor strategies from neoliberalism to protectionism. She argues that for strategies to be pro-poor, growth must deliver disproportionate benefits to the poor to reduce inequalities, which have limited the potential for poverty alleviation. She argues that there is a need to shift policy from growth to equity; however, that would involve a shift away from the neoliberal approach, and therefore a policy based on equity is more likely to remain as rhetoric rather than be implemented on a large-scale basis.

Volunteer tourism

The above sections on pro-poor tourism, fair trade and NGOs reflect to a large degree on what policies initiated by governments, NGOs or the industry can do to assist local communities. Volunteer tourism takes the focus to the tourists themselves. Wearing (2001: 1) states that volunteer tourism 'applies to those tourists who volunteer in an organised way to undertake holidays that might involve aiding or alleviating the material poverty of some groups in society, the restoration of certain environments or research into aspects of society or environment'. While the tourist provides some assistance in some form, the benefits, according to Wearing (2001), not only go to the host community but the experience also benefits the tourists. The experience can cause value change and altered consciousness in tourists, subsequently changing their lifestyle. Scheyvens (2002) describes volunteer tourism as an element of 'justice tourism' with individuals paying to travel to developing countries to assist with development or conservation projects. Wearing (2001) also states that volunteer tourism may be viewed as a development strategy leading to sustainable development. Conservation work is carried out in a variety of destinations such as Africa, Asia, Central and South America, and can occur in places such as rain forests, cloud forests, conservation areas and biological reserves (Wearing 2001). Campbell and Smith (2006) examined the values of volunteer tourists working for sea turtle conservation in Tortuguero, Costa Rica. The volunteering is organized by the Caribbean Conservation Corporation (CCC), a NGO headquartered in the United States but with a national office in Costa Rica, and a year-round field station in Tortuguero. The second main type of volunteer tourism is linked to development work by tourists. These types of projects can involve offering medical assistance, projects linked to heritage and cultural restoration, and other types of social and economic development initiatives (Wearing 2001). In the case of South Africa, Stoddart and Rogerson (2004) examined volunteer tourists volunteering with Habitat for Humanity International's Global Village Work Camp Programme. Volunteers work with local community members, thereby raising their own awareness of poverty and building decent, affordable housing.

One segment that receives a fair bit of attention is students volunteering during their 'gap year', which tends to be between finishing school and starting university. This is very popular, especially in Britain, although Simpson (2004) indicates that the demographics of the 'gap year' are expanding to people taking 'career breaks'. There are numerous Internet sites advertising various 'gap year' experiences. The phenomenon of youth travelling for substantial periods of time is not limited only to Britain; young people from countries such as Australia and New Zealand are also known for longer trips abroad. While volunteer tourism has the potential to assist communities, concerns have also been raised. Simpson (2004: 690) states:

> currently, the gap year industry promotes an image of a 'third world other' that is dominated by simplistic binaries of 'us and them', and is expressed through essentialist clichés, where the public face of development is one dominated by the value of western 'good intentions'.

Other concerns raised include whether the volunteers are actually making a difference or causing more difficulties than providing benefits, and whether the gap year is a new form of colonialism (Brown 2006).

Gender and community development

A theme of this chapter is that communities are made up of individuals and groups; they do not all have the same values and goals, and they do not face the same challenges or have the same opportunities. Within the United Nations Millennium Development goals both women and children are specifically mentioned. In India, for example, a law recently came into effect banning children under 14 years of age from working as servants in homes, or in restaurants, tea-shops, hotels and spas (Associated Press 2006). There is a discourse that highlights the links between the social and economic position of women in developing countries and environmental degradation (Baker 2006: 167):

> The position of women makes them more vulnerable to the negative effects of environmental degradation than their male counterparts. They are more marginalised, usually work harder, especially if engaged in agricultural labour, have a less adequate diet and are often denied a voice in the political, economic and social spheres.

In the tourism impact literature, the relationship between women and tourism has been discussed from a number of different perspectives, including women having more opportunities as a result of tourism as well as being exploited by tourism such as working in the sex industry (Hashimoto 2002). Studies have also identified that it is often the managerial positions that are taken by men, while women are more likely to occupy lower payed, part-time and seasonal positions (Wall and Mathieson 2006).

A change has occurred within the women and environment debate as the discussion has shifted to sustainable development (Baker 2006). The early discussions focused on women as passive victims of environmental degradation stemming from global processes and, more recently, there has been increased emphasis on women's positive roles as efficient environmental resource managers in developing countries. Baker (2006: 168) highlights some ways in which women can be promoters of sustainable development:

- With their domestic, agricultural and cultural roles, women are key agents in promoting sustainable development.
- Women hold knowledge of their local environment and they are the key to developing appropriate biodiversity strategies.
- Promoting sustainable livelihoods at community level can be accelerated by giving women the right to inherit land as well as to have access to credit and resources.
- A strong connection exists between promoting human rights, especially for women, and promoting sustainable development, as it is based on equity and partnership.

- The promotion of democratic environmental governance has a gender dimension, as participation based on gender equality of access is more democratic, legitimate and effective.

The above list illustrates some of the key areas where women can be promoters of sustainable development in their communities. These roles are of major importance in the context of tourism. In community-based tourism, for example, women have the potential to help promote sustainable development in their interactions with the tourists and the natural environment. The tourism industry has created new opportunities for women, particularly in developing countries and rural areas (Wall and Mathieson 2006), with a variety of positions in both the formal (e.g. hotels and running guest-houses) and the informal sector (e.g. guides and street/beach vendors) (see Plate 5.6). It has also generated greater independence and more income for some (Hashimoto 2002; Momsen 2004).

Plate 5.6 ***Indonesia, Lombok: Young women present traditional Sasak weaving.***

Conclusion

Preston (1996) suggests that there are three main paradigms in terms of characterizing and securing development. The first paradigm focuses on state intervention, the second on the role of the free market, and the third relates to the power of the political community. This chapter largely focuses on the third aspect: the power of the political community and the development of tourism at the local level. As Preston (1996) suggests, this approach is oriented towards securing formal and substantive democracy and has an institutional vehicle through NGOs, charities and dissenting social movements. In the context of tourism it is important to recognize to what degree local communities interact with the state and the market forces. A key issue with respect to the relationship between tourism and communities is the power relationships that exist within the community and between communities and the tourism sector. Does the local community have a chance to participate in a meaningful way? As Milne and Ewing (2004: 215) state, the critical issue becomes 'how do we ensure not only that local involvement and participation occurred, but that it can be sustained in such a way that it leads to effective development outcomes?' The development benefits of tourism need to extend to as many as possible in the community. Critics focusing on the power of the political community have highlighted the alleged impractical idealism of the approach along with the potential to generate conflict between powerful groups in the peripheries and metropolitan centres which can make the situation of the poor worse (Preston 1996). Others have suggested that problems of consensus building, barriers to participation, lack of accountability, weak institutions, and a lack of integration with international funding organizations are barriers to indigenous or community development (Wiarda 1988; Brinkerhoff and Ingle 1989).

Communities are complex, and are increasingly recognized as important resources for tourism. With the shift to sustainability, participation of local communities and capacity building has become central to many tourism plans. Communities are not only impacted by tourism but they also respond and take advantage of the opportunities from tourism (Wall and Mathieson 2006). This chapter has referred to several types of tourism as well as several strategies that can be adopted to promote additional benefits to disadvantaged communities. Caution, however, needs to be used when comparing different types of tourism. In the tourism literature comparisons have been made between large-scale mass tourism and small-scale community-based tourism. As Butler (1993: 34) suggested, 'making simplistic and idealised comparisons of hard and soft, or mass and green tourism, such that one is obviously undesirable and the other close to perfection, is not only inadequate, it is grossly misleading'. He goes on to state that mass tourism does not have to be uncontrolled, unplanned, short term or unstable; green tourism – or in the case of this chapter we could add community-based tourism – is not always inevitably considerate, optimizing, controlled, planned and under local control. Community-based tourism may well generate conflict, inequities and resource exploitation. The economic and political realities of the marketplace with respect to tourism and the communities where tourism occurs need to be considered. A highly fragmented and extremely competitive tourism industry in the public and private sectors often mitigates against self or internal control (Butler 1993). Having examined the nature of 'host' communities and various strategies of community-based tourism, the next chapter will focus on the 'guest' by exploring the consumption of tourism.

Discussion questions

1. **Distinguish the strengths and weaknesses of both the informal and formal tourism employment sectors.**
2. **Can community-based tourism development help promote community and individual empowerment?**
3. **Does hosting international tourists reinforce notions of dependency for communities?**
4. **Does community-based tourism represent a form of sustainable tourism?**
5. **What role should NGOs take in community-based tourism?**

Further reading

Richards, G. and Hall, D. (eds) (2000) *Tourism and Sustainable Community Development*, London: Routledge. [This edited volume of 20 chapters has a range of international case studies and explores what local communities can do to contribute towards sustainable tourism, and what sustainability is able to offer local communities.]

Singh S., Timothy, D. and Dowling, R. (eds) (2003) *Tourism in Destination Communities*, Wallingford: Oxon: CABI. [This edited volume contains 14 chapters divided into three sections that explore the relationships between tourism and the destination community, the impacts of tourism on destination communities, and the challenges and opportunities for destination communities.]

Websites

This website illustrates how the United Nations works with civil society organizations: www.un.org/issues/civilsociety/.

This United Nations web page has links to information and resources on gender equality and empowerment of women: www.un.org/womenwatch/.

6 The consumption of tourism

Learning objectives

When you have finished reading this chapter, you should be able to:

- **Appreciate the factors that influence tourist consumer behaviour;**
- **Understand the changing nature of tourism demand;**
- **Identify the extent of emerging, new, environmentally aware, tourism markets;**
- **Identify ways in which the tourism industry can influence tourist behaviour;**
- **Appreciate the importance of domestic tourism in development.**

Tourism is, essentially, a social activity. Certainly it is big business and one of the world's largest economic sectors, providing a significant source of income and foreign exchange earnings for many countries; certainly it is a vast and diverse industry, providing employment for up to 10 per cent of the global workforce; and certainly, as this book explores, it is a potentially effective agent of development. However, it is important not to lose sight of the fact that, first and foremost, tourism is about people – it is about millions of individuals travelling within their own countries or overseas, visiting places and attractions, staying in destinations and engaging in various activities during their stay.

More specifically, tourism is about people – tourists – who interact with and impact upon other people and other places. In other words, a fundamental characteristic of tourism is that the product is 'consumed' on site; whether in their home country or overseas, tourists must travel to the destination to enjoy or participate in tourism. In fact, as with all services, the tourism product – normally thought of as tourist experiences – cannot actually be produced without the input of tourists into the production process. Although hotels, restaurants, shops, transport operators, attractions and other businesses within the tourism sector offer potential services, these services (or experiences) are not provided until tourists actually purchase and consume them (Smith 1994). Thus, tourism development inevitably results in the presence of tourists in the destination; equally, the presence of tourists inevitably brings about consequences (both positive and negative) for the destination's environment, its economy and local communities.

The consequences or impacts of tourism development are explored in more detail in the next chapter. However, it has long been recognized that the nature of these impacts cannot be divorced from the nature of the consumption of tourism itself. Although the emergence of mass international tourism in the 1960s was greeted with some degree of optimism, the potential economic benefits of tourism being seen as a panacea to the developmental challenges facing destinations and their communities, it was not long before concern was being expressed about the adverse effects of mass tourism consumption on local environments and cultures. One commentator at that time, for example, observed that, as mass tourism develops:

> local life and industry shrivel, hospitality vanishes, and indigenous populations drift into a quasi-parasitic way of life catering with contemptuous servility to the unsophisticated multitude.
>
> (Mishan 1969: 142)

During the 1970s, books such as *Tourism: Blessing or Blight?* (Young 1973) and *The Golden Hordes: International Tourism and the Pleasure Periphery* (Turner and Ash 1975) adopted a more balanced perspective; yet the overall tone was that mass tourism may have consequences for the destination that potentially outweigh the developmental benefits of tourism. Indeed, by the 1990s, most if not all of the problems associated with tourism development were being blamed on mass tourism and tourists:

> the crisis of the tourism industry is a crisis of mass tourism; for it is mass tourism that has brought social, cultural, economic and environmental havoc in its wake, and it is mass tourism practices that must be radically changed to bring in the new.
>
> (Poon 1993: 3)

The 'new' proposed by Poon refers to alternative (to mass), sustainable forms of tourism development, the achievement of which, as noted in Chapter 2, is dependent on the adoption of a new 'social paradigm' with respect to the consumption of tourism. In other words, fundamental to the successful development of more appropriate, sustainable forms of tourism is the need for tourists to act more sustainably, to become 'good' or 'responsible' tourists. Poon also suggested that, by the early 1990s, there was, in fact, evidence of the emergence of 'new' tourists – that is, tourists who are more environmentally aware, more quality conscious, more adventurous, and more ready to reject the passive, structured, mass-produced package holiday in favour of more individualistic, authentic experiences. Since then, not only has it become widely assumed that tourists are becoming 'greener' or increasingly disposed towards consuming tourism in more responsible or environmentally appropriate ways, but also that assumption has frequently been used as the justification for developing or promoting sustainable forms of tourism.

Importantly, however, there is little evidence to support this claim that tourists are in fact becoming greener. For example, it is often suggested that the growth in demand for ecotourism is a sign that tourists are increasingly seeking environmentally appropriate tourism experiences. It is certainly true that ecotourism is becoming increasingly popular; indeed, recent research suggests that the number of tourists taking ecotourist holidays is growing three times faster than those choosing 'mainstream' holidays and that, by 2024, ecotourism will represent 5 per cent of the global holiday market (Starmer-Smith 2004). However, significant doubt exists about the extent to which 'ecotourists' are motivated by genuine environmental concerns (Sharpley 2006b). Similarly, while surveys suggest that many tourists would be willing to pay more for environmentally friendly holidays (see e.g. Tearfund 2000), there is limited evidence that this is manifested in practice.

The purpose of this chapter, therefore, is to explore the nature of the consumption of tourism and the consequential implications for sustainable tourism development. The first questions to address briefly are: What is the tourism demand/consumption process? What are the factors that influence that process? How has the consumption of tourism changed over time?

The tourism demand process

Tourism demand or consumption is a complex process, described by one commentator as 'discretionary, episodic, future oriented, dynamic, socially influenced and evolving' (Pearce 1992: 114). In other words, the demand for tourism involves making choices about how to spend specific periods of leisure time, choices which may be influenced by a variety of factors and which may change over time. Moreover, it is not only about how and why people decide to participate in tourism, but also about how they behave as tourists, why they choose particular types of tourism, what tourism means to them, and why their 'tastes' in tourism may change.

Despite this complexity, however, tourism demand is seen typically, though somewhat simplistically, as a sequential set of stages which may be summarized as follows:

Stage 1 Problem identification/felt need.
Stage 2 Information search and evaluation.
Stage 3 Purchase (travel) decision.
Stage 4 Travel experience.
Stage 5 Experience evaluation.

Each stage in the demand process may be influenced by personal and external variables, such as time and money constraints, social stimuli, media influences, images/perceptions of the destination, or marketing, while each consumption experience feeds into subsequent decision-making processes (Figure 6.1).

At the same time, of course, tourism demand is not a 'one-off' event. People consume tourism over a lifetime, during which tourists may climb a 'travel career ladder' (Pearce 1992) as they become more experienced tourists. As a result,

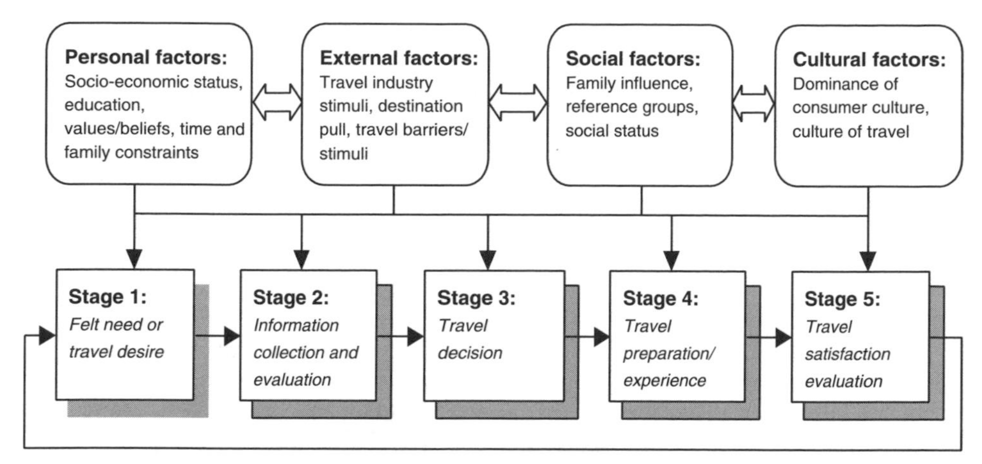

Figure 6.1 ***The tourism demand process.***

tourists' travel needs and expectations may change and evolve, but these may also be framed and influenced by evolving social relationships, lifestyle factors and constraints, and emerging values and attitudes. This latter point is of particular relevance to the theme of this chapter and will be returned to shortly, but fundamental to understanding the consumption of tourism is an appreciation of the factors that may influence the demand process. These may be categorized under four headings (see Cooper *et al.* 2005: 53):

1 *Energizers of demand.* These are the forces and influences (or personal 'push' factors) that collectively create the motivation to travel or go on holiday, or initiate the demand process.
2 *Effectors of demand.* The information search/evaluation process and subsequent purchase decision is influenced by the tourist's knowledge and perceptions of particular places, destinations or experiences. These are sometimes referred to as destinational 'pull' factors which lead the tourist to making particular travel choices.
3 *Filterers/determinants of demand.* A variety of economic, social and demographic factors determine particular choices or 'filter out' inappropriate products. These include: mobility; employment and income; paid holiday entitlement; education levels; and age, gender, race and stage in family life cycle. In addition, choice may be determined by intangible, psychographic variables, such as attitudes, values and lifestyle.
4 *Roles.* Holiday/travel choices are also influenced by roles within the purchasing 'unit' (for example, the different roles adopted by family members in choosing a holiday) and as tourists.

Evidently, then, an almost infinite combination of variables may influence how, when and where tourism is consumed by individual tourists, to the extent that predicting tourist behaviour may be seen as a difficult if not impossible task. In fact, according to Krippendorf (1987), many tourists themselves are unable to explain precisely why they participate in particular types of tourism, while others, such as Ryan (1997), suggest that tourism is an irrational (and, hence, inexplicable) form of behaviour! Nevertheless, for the purposes of this chapter, it is important to consider three issues or factors that influence the tourism demand process, namely: tourist motivation; the influence of values on consumption; and tourism and consumer culture.

Tourist motivation

Academics and researchers have long been concerned with the questions of why and how people consume or participate in tourism. In particular, attention has been (and continues to be) focused on the first stage in the tourism demand process, or what is broadly referred to as tourist motivation. The reason for this is, perhaps, self-evident. Not only is motivation the 'trigger that sets off all events in travel' (Parinello 1993), the process that translates a felt need into goal-oriented behaviour (i.e. tourism) directed at satisfying that need, but also the nature of the felt need influences the consequential behaviour of the tourists and their potential impact on destinations.

Given the complexity of the subject, it is not surprising that tourist motivation is explored from a variety of perspectives (Sharpley 2003). The key themes that emerge from these perspectives are summarized below, but first it is useful to consider so-called 'tourist typologies' as descriptors of distinctive forms of tourist consumer behaviour.

Tourist typologies

Tourist typologies are, in essence, lists or categorizations of tourists based on a particular theoretical or conceptual foundation. As such, they tend to be descriptive as opposed to predictive, yet they do reflect, if not explain, different motivations, interests and styles of travel on the part of tourists. One of the first such typologies was proposed by Gray (1970) who coined the terms 'sunlust' and 'wanderlust', where sunlust tourism is essentially resort-based and motivated by the desire for the three Ss – sun, sea and sand. Conversely, wanderlust tourism is typified by the desire to travel and to experience different places, peoples and cultures. Implicit in each term are the characteristics of the different forms of travel and the potential destinational impacts of each.

The distinction between these two types of tourism was expanded on by Cohen (1972) in his widely cited tourist typology based on a 'familiarity–strangerhood' continuum. In other words, Cohen suggested that tourists are more or less willing to seek out different or novel places and experiences; some travel within an 'environmental bubble' of familiarity – they seek out the normal/familiar (food, language, accommodation, fellow tourists) and are unwilling to risk something new or different – whereas others seek out different or unusual experiences. This, in turn, determines how different tourists travel. Some are 'institutionalized' inasmuch as they depend upon the tourism industry to provide familiar, predictable, organized and packaged holidays; others, conversely, are 'non-institutionalized',

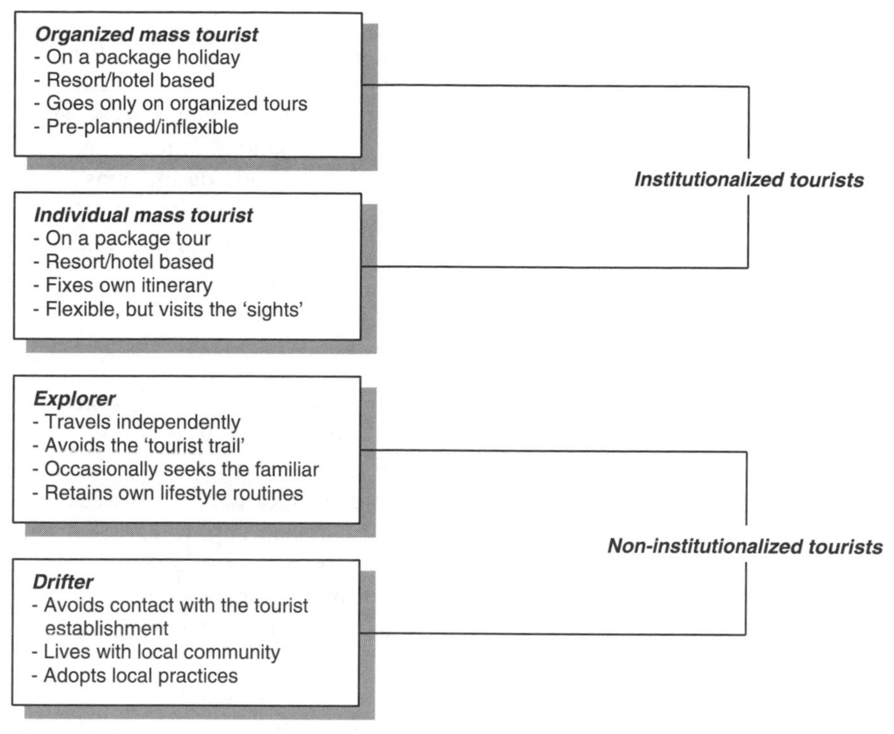

Figure 6.2 ***Cohen's typology of tourists (1972).***

travelling independently and requiring little contact with the tourism establishment. Cohen's typology proposes four types of tourist, from the organized mass tourist to the drifter (Figure 6.2).

The extent to which this typology remains relevant to contemporary international tourism is debatable. Not only has the practice of tourism matured over the past three decades – as discussed later in this chapter, the demand for tourism in terms of places, experiences and frequency has changed significantly, as has the way in which people actually organize and purchase their trips or holidays – but also the phenomenon of globalization has transformed the world within which tourism occurs. In particular, global communications and information systems have made it virtually impossible (for explorers or drifters) to escape the realities of their 'home' world, while the practice of independent travel has, in effect, become institutionalized. Nevertheless, it provides a useful framework for thinking about how different types of tourists behave and interact with their destinations.

The same may be said for a number of other typologies. Cohen (1979), for example, subsequently constructed a 'phenomenology of tourist experiences', based again on the familiarity–strangerhood continuum. However, the focus was on the extent to which tourists either enjoy a sense of belonging or feel like strangers in their normal home environment. Those who benefit from a satisfying, fulfilling home life are, according to Cohen, likely to look for only trivial, recreational experiences on holiday while, at the other extreme, 'alienated' individuals are likely to seek out more meaningful or authentic experiences when on holiday. Alternatively, Plog's (1977) widely cited (though frequently criticized) typology attempted to link tourists' personality with destination choice; risk-averse (psychocentric) tourists travel to familiar, closer destinations, while more adventurous risk takers (allocentrics) choose more distant, exotic destinations. Again, however, transformations in the supply of tourism, particularly the expansion of the 'packaged' pleasure periphery, limit the contemporary relevance of this typology.

Smith's (1989) typology deserves mention in that it explicitly links tourist types and numbers to destinational impacts. At one extreme, small numbers of 'explorers' travel to undiscovered places and, Smith suggests, attempt to blend into the destination; at the other extreme, large numbers of charter tourists care little for the destination and have maximum negative impact. Within these two extremes, different types/numbers of tourists travel with different attitudes towards the destination. However, as with most typologies, a causal link between tourist type, tourist behaviour and subsequent destinational impact is, often mistakenly, assumed. For example, in many destinations in the less developed world, mass-institutionalized tourism exists in the form of all-inclusive resorts; yet, in some cases, these may bring substantial economic benefits but only limited environmental impact (see Box 6.1).

More recently, Yiannakis and Gibson (1992) developed a typology that identified 15 tourist roles, from 'sun lover' to 'educational tourist', each determined by a preferred balance between familiarity–strangeness, stimulation–tranquility and structure–independence. Their work was subsequently 'tested' in an Australian context, confirming that tourists choose destinations according to the extent that they will be able to enact their preferred roles as tourists (Foo *et al.* 2004).

Typically, however, tourist typologies tend to imply that larger groups of mass, 'institutionalized' tourists on organized holidays and seeking more traditional sun-sea-sand holidays have a greater negative impact on destination environments and societies than do smaller numbers of independently minded, environmentally aware tourists seeking more individualistic and cultural experiences. However, as the following section on tourist motivation suggests, care must be taken in making such a broad assumption.

Tourist motivation: key themes

As noted above, tourist motivation is fundamental to the consumption of tourism. Not only does the motivational stage 'kick-start' the entire tourism demand process, but also the manner in which people behave as tourists (the experiences they seek as well as their behaviour in a more specific sense) is very much determined by what motivates them in the first place. Therefore, understanding tourist motivation is fundamental to understanding the overall consumption of tourism.

There is enormous diversity in the treatment of tourist motivation within the tourism literature, reflecting both the variety of ways in which it is possible to

Box 6.1

All-inclusive resorts – the case of Sandals, Jamaica

The concept of all-inclusive international holiday resorts can be traced back to the development of Club Méditerranée (Club Med) resorts in the 1950s. However, the opening of the first Sandals resort at Montego Bay, Jamaica, in 1981 marked the beginning of a new era in all-inclusive, club-based holidays. Not only was the resort truly all-inclusive – virtually every aspect of the holiday was prepaid – but it also provided a high-quality experience for couples only. Since then, Sandals has expanded its operations across the Caribbean, with a total of 17 resorts in Jamaica, Cuba, The Bahamas, St Lucia and Antigua, providing luxury holidays for a variety of niche markets. In fact, the Caribbean has become widely associated with all-inclusive holidays, which are now the most popular form of international tourism in the region; in addition to Sandals, SuperClubs and Allegro Resorts also own a large number of resorts around the Caribbean, and it has been claimed that 48 of the world's top one hundred all-inclusive resorts are located in the Caribbean.

All-inclusive resorts have long been criticized for limiting the 'trickle-down' economic benefits of tourism, for a high level of imports, for restricting employment and promotion potential and, generally, for providing little developmental benefit to the destination. However, in addition to meeting substantial demand for all-inclusive holidays (i.e. tourists who seek safe, luxurious holidays in an exotic location), the resorts in Jamaica collectively generate the largest contribution to GDP on the island while, in terms of employment, they create relatively more jobs than do conventional hotels. For example, research has shown that five-star all-inclusives create between 1.5 and 2 jobs per room compared with one job per room in a conventional hotel. At Sandals in particular, staff are relatively well paid and benefit from free meals and transport; it is claimed they can save up to one-third of their monthly salary. All line staff receive at least 120 hours of training each year and, in 2003, Sandals Montego Bay initiated a skills training centre for local school leavers. In addition, Sandals established a Farmers Programme in 1996, under which local farmers were encouraged to supply produce to the resorts. By 2004, 80 farmers were supplying hotels across the island and total annual sales had risen to US $3.3 million. Local craft producers are also able to sell their products to tourists within the resorts while, finally, it is important to note that, with the exception of those in Cuba, all Sandals resorts are owned by the original Jamaican company, limiting the problem of profit repatriation associated with other international hotel chains.

Sources: Issa and Jayawardena (2003); Ahley *et al.* (2006).

approach the subject and the fact that a generally accepted theory or understanding of tourist motivation has yet to emerge. Nevertheless, a number of key points or themes are in evidence, the first of which is the fact that motivation should not be equated with demand; that is, tourism demand or consumption is the outcome of motivation. To put it another way, tourist motivation may be seen as a 'meaningful state of mind which adequately disposes an actor or group of actors to travel, and which is subsequently interpretable by others as a valid explanation for such a decision' (Dann 1981). Therefore, what we are concerned with essentially are the factors or influences that create this 'meaningful state of mind'. Such influences may be either intrinsic or extrinsic to the individual tourist.

Intrinsic motivation

The study of motivation has traditionally been concerned with people's innate needs and desires; it is an individual's intrinsic and deep-rooted needs that lead to motivated, goal-oriented behaviour. In other words, every individual has unique personal needs, the satisfaction of which has long been considered the primary arousal factor in motivated behaviour. In this context, one of the best-known and most widely applied theories is Maslow's hierarchy of needs; in fact, many tourism texts refer to Maslow's model, linking specific needs with identified goal-oriented tourist behaviour, while others explicitly adapt it, as, for example, in the case of the travel career concept mentioned above. Similarly, Crompton (1979) suggested that tourist motivation emanates from the need to restore an individual's psychological equilibrium which may become unbalanced as a result of unsatisfied personal needs. However, most commentators accept that tourist motivation is not simply a function of intrinsic, psychological needs – extrinsic or social factors are also significant influences on why and how people consume tourism.

Extrinsic motivation

From an extrinsic perspective, tourist motivation is structured by the nature and characteristics of the society to which the tourist belongs. To put it another way, there are a variety of forces and pressures in an individual's social and cultural environment which may influence that individual's needs and motivations, or desire to consume tourism. At a basic level, for example, the motivation for many people to take a holiday is to relax, to rest, to have a change and to get away from the routine; in order to survive in modern society an individual must periodically escape from it. Indeed, it is somewhat ironic that modern society has created both the need for and means of mass international travel. That is, technology, economic growth and socially sanctioned free time have brought freedom and mobility to many in the developed world; yet, in modern society, tourism has become an essential 'social therapy, a safety valve keeping the everyday world in good working order' (Krippendorf 1986).

Importantly, it is not simply the desire to get away that may motivate people. Tourism also offers the opportunity to travel beyond not only the physical, geographic boundaries of normal, home existence, but also the social boundaries. In other words, tourism releases people from the constraints or social 'norms' of their home existence, providing them with the opportunity to indulge in so-called 'ludic' behaviour (that is, behaviour that may be described as play), manifested in, for example, excessive drinking or sexual activity or more generally 'letting their hair down'. Conversely, tourists may seek meaning or 'authenticity' through tourism, compensating for what may be regarded as the lack of meaning or 'reality' in modern society. Indeed, MacCannell (1989) views the tourist's search for meaning and authenticity as a modern form of pilgrimage, the motivation for tourism thus becoming a form of secular spiritual quest.

Beyond the more general relationship between the nature of modern social life and tourist motivation, a number of other social influences may also be identified. Some commentators explore the relationship between work and leisure/tourism experiences (Ryan 1991), the nature of work having a potential impact on the desired form of tourism consumption. For example, it is suggested that those in demanding, stressful jobs may seek compensatory (relaxing, restful) holidays, and vice versa. Others refer more generally to cultural and social factors, including social class, reference groups and family roles, as dominant social influences on tourist motivation and behaviour.

Collectively, this diversity of intrinsic and extrinsic motivating factors suggests that identifying specific or dominant determinant factors may be a difficult if not impossible task, particularly given the fact that tourists themselves may be unwilling or unable to express their real travel motives. Nevertheless, there are two generally accepted features of tourist motivation:

- *Tourism as escape*. Whether implicitly or explicitly, tourists are motivated primarily by their desire to escape, to travel away from rather than to travel to somewhere or something. Reviews of the relevant literature consistently find that the escape motive predominates.
- *Tourism as 'ego-enhancement'*. Tourists are motivated by the potential rewards of participating in tourism. Such rewards may be personal, interpersonal, psychological or physical and, collectively described as 'ego-enhancement', they compensate for the deficiencies or pressures and strains of everyday life. At the same time, tourists' motivations are markedly self-oriented or egocentric, focusing on personal wants and needs. In other words, tourism represents a form of self-reward or self-indulgence.

The implication of this is, of course, that tourism is essentially an egocentric, 'selfish', escapist activity. It is, for the most part, about relaxation, fun and entertainment, and hence tourists are most likely to give priority to satisfying their personal needs rather than demonstrating and responding to a positive concern for the consequences of their actions – their focus will be inward, on the satisfaction of personal needs and wants, rather than on the external tourism environment. This is not to say that the 'green' tourist does not exist in reality, but the study of tourist motivation certainly confirms McKercher's assertion, referred to in Chapter 2, that tourists are simply consumers, not anthropologists wishing to blend into and learn about local communities and cultures. This debate is addressed in more detail later in this chapter but, as suggested earlier, both values and consumer culture must also be considered as influences on tourism consumption.

Values and tourism consumption

Psychographic variables, such as values, attitudes and opinions, are important determinants in the tourism decision-making process. In particular, values, which serve as standards or criteria for all forms of personally or socially preferable behaviour, have been found to be potential predictors of travel behaviour. In other words, how people consume tourism is likely to be influenced strongly by their personal values. For example, if family security is highly valued, then safe and predictable family holidays may be the preferred tourism experience. Conversely, a backpacker or independent traveller may have an exciting life as a dominant value.

Importantly, values are usually organized into a value system or hierarchy. That is, individuals typically hold a number of values which may be more or less influential in different contexts, while different people attach varying values to similar objects or behaviour; 'an individual relies on his/her value system to maintain self-esteem or consistency in those situations where one or more conflicting values are activated' (Madrigal and Kahle 1994). In the tourism context the likelihood of value conflict is high, particularly between personal values such as pleasure, freedom or happiness, and social values that serve as guidelines for socially acceptable behaviour. Implicitly, therefore, only individuals who hold strong or dominant environmental values are likely to adopt a 'responsible' approach to the consumption of tourism, particularly given the motivating factors identified in the preceding section.

Tourism and consumer culture

The discussion thus far has focused specifically on tourism. However, tourism is just one of a whole host of goods and services that people in modern societies consume and, as Solomon (1994: 536) observes, 'consumption choices simply cannot be understood without considering the cultural context in which they are made'. Therefore, it is important to consider briefly how the consumption of tourism is framed by a wider consumer culture.

In modern tourism-generating societies, consumption has become a culturally significant activity. For modern consumers, goods, services and experiences such as tourism have assumed a meaning and significance well beyond their simple, utilitarian values, and the practice of consumption plays an important role in contemporary social life (Lury 1996). In other words, all goods and services satisfy basic needs – food to satisfy hunger, clothes to keep warm, holidays for rest and relaxation and so on – but their consumption also fulfils a wider role.

In particular, consumption is seen as an effective means of social classification, creating self-identity or social status. This is not, of course, a new phenomenon – so-called 'conspicuous consumption' has, in general, long been a marker of wealth and social status – while tourism in particular has always been used as an indicator of status, wealth or taste. However, in modern societies, consumption has become a dominant and widespread element of social life, with people increasingly seeking identity or distinction through the products they consume. As a result, producers in general have had to become increasingly responsive to the diverse and rapidly changing demands of consumers; in tourism, this has been manifested in, for example, the development of specialized, niche products or styles of tourism which, though relatively affordable, have the aura or status of luxury. For example, ecotourism, which tends to be exclusive (expensive) tourism based in exotic or distant places, is often referred to as 'ego-tourism' (Wheeller 1992).

In addition to classification purposes, products and experiences may be consumed in other ways. It has been suggested, for example, that three further modes of consumption exist (Holt 1995):

- *Consumption as play*. Consumption that focuses on interaction with other consumers rather than on the object of consumption, such as the shared experience of 'white-knuckle' rides at a theme park.
- *Consumption as integration*. The integration of the self into the object of consumption or vice versa, whereby tourists may adapt themselves to the destination environment and culture (i.e. become 'good' tourists), or participate in activities that reflect their self-image, such as trekking holidays if they consider themselves to be adventurous.
- *Consumption as experience*. Consumption that is framed by a social world that provides meaning to the object of consumption. For tourists, the social meaning of tourism (for example, opportunities for hedonistic pleasure) defines their tourism consumption.

Quite evidently, then, the demand for tourism is a highly complex issue, the study of which embraces a variety of themes and perspectives which this brief overview has only been able to touch on. Nevertheless, it has demonstrated that the consumption of tourism, or why and how people participate in tourism, is influenced by an almost infinite variety of personal (intrinsic) and socio-cultural (extrinsic) factors. At the same time, not only do people become more experienced tourists at they progress through their 'travel career' and adapt their behaviour as their values, attitudes and personal circumstances change, but also the overall practice of tourism consumption is constantly evolving. Therefore, it is a difficult if not impossible task to predict individual tourist behaviour, although, as the following section shows, a number of distinctive trends exist in the demand for tourism which are relevant to the management and development of tourism.

Tourism demand: trends and changes

As noted in Chapter 1, there are two principal ways of exploring trends and changes in the demand for tourism: historical and contemporary flows in international tourism (i.e. spatial transformations evidenced by statistical data); and transformations in the nature of tourism demand (i.e. changes in the style of tourism demand). Both perspectives are important in the context of this book. First, contemporary trends in the volume, value and direction of international tourism flows provide an indication of tourism's increasing contribution to development in emerging destination areas. More specifically, international tourist flows have traditionally been polarized and regionalized between and within the wealthier, developed regions of the world. However, in recent years, many less developed countries have been claiming a greater share of global tourist arrivals and receipts and, consequently, have benefited from tourism-related economic growth and development. Second, changes in the nature of demand point to both opportunities and challenges in managing tourism development effectively. For example, new niche products, such as adventure tourism (see Plate 6.1), ecotourism or heritage tourism, represent opportunities for tourism development in countries with a rich natural and cultural heritage; equally, the increasing popularity of all-inclusive holidays has underpinned the successful development of tourism in a

Plate 6.1 ***Tunisia, near Matmata: These camels are used to provide rides to tourists.***

number of less developed countries. For example, the rapid and successful development of tourism in Cuba since the late 1980s (tourism is now the island's principal source of hard currency earnings) has been based primarily on all-inclusive resorts (Martin de Holan and Phillips 1997; Cerviño and Cubillo 2005).

International tourism flows are considered in some detail in Chapter 1. The two key trends highlighted are as follows:

- *Global growth*. Over the past half century, worldwide tourist arrivals and receipts have demonstrated consistent overall growth, pointing to tourism's resilience to external 'shocks', such as economic recession, wars, health scares and so on. Such growth continues unabated; despite the Asian tsunami, oil price rises, health scares and terrorism, worldwide arrivals in 2005 totalled 808 million, representing a 5.5 per cent increase on the previous year (WTO 2006a). It is forecast that, by 2020, this figure will have almost doubled to 1.5 billion international arrivals annually. It is also important not to overlook domestic tourism which, globally, is thought to be between six and ten times greater in terms of volume/trips. Accurate data do not exist but, for some less developed nations, domestic tourism remains far more significant than international tourism. In India, for example, international arrivals amounting to 2.6 million in 2000 were dwarfed by an estimated 320 million domestic trips.
- *Global spread*. While Europe has maintained its dominant share of international arrivals, that share has been declining. Conversely, since 2000,

the highest average annual growth rate (9 per cent) in arrivals has been experienced by the Middle East region, followed by East Asia and Pacific (7 per cent) and Africa (5 per cent); over the same period, North America experienced a 2.1 per cent fall in arrivals (WTO 2006a). This is evidence both of tourists from the main generating countries in Europe and North America travelling to more distant or exotic destinations and, more particularly, of increased intra-regional travel – that is, much of the growth in tourism in areas such as the Middle East and Asia has been generated within those regions, fuelled by rapid economic growth. Such an increase in intra-regional travel is likely to continue apace, particularly given the rapid economic development in China and India.

Changes in the nature of tourism demand

Although the extent to which the 'new' tourist, referred to earlier in this chapter, exists in reality (or, more precisely, the extent to which tourists are adopting a more responsible attitude towards their consumption of tourism) is debatable, there is no doubt that styles of tourism consumption have undergone significant transformations over the past half century, and particularly over the past two decades. Such transformations relate to: types of holidays, with a rapid proliferation in differentiated, niche products; how such holidays are organized and purchased; and an increasingly fuzzy relationship between tourist markets and tourism products. Collectively, they represent a shift from the standardized two-week, short-haul, sun-sea-sand package holiday to more individualistic, active/participatory forms of tourism that provide a wider variety of physical, cultural or educational experiences. Nevertheless, the traditional 'package' remains popular; in 2001, for example, over 20 million overseas trips by UK residents were on inclusive/package holidays (IRN 2002).

It is not possible here to explore all these changes in detail. Not only are there innumerable categories of new tourism products or experiences, which are frequently subcategorized (for example, adventure tourism is subdivided into 'hard' and 'soft' adventure tourism), but also a lack of definitional distinction exists between them. For instance, terms such as 'wildlife tourism', 'wilderness tourism', 'nature tourism' and even 'ecotourism' are frequently used interchangeably. As a result, accurate data regarding participation in particular forms of tourism do not always exist. Nevertheless, for our purposes it is useful to highlight the general trends occurring under the three interrelated headings referred to above.

Types of tourism/holidays

The principal trend in the demand for tourism over the past two decades has been the dramatic increase in the demand for specialized, niche products as an alternative to the traditional summer-sun holiday. Of course, other forms of tourism have long been offered by the tourism industry, such as winter sports tourism or winter sun tourism (both of which were initially upmarket forms of tourism but soon became more mainstream products), or cultural/educational tourism. However, there has been a rapid increase in the supply of niche tourism products designed to meet the needs of tourists with special interests, who wish to do something different, or who perhaps want to be seen doing something different, exclusive or exotic. Such niche products are frequently labelled under generic headings, such as adventure tourism, cultural tourism, nature tourism, heritage tourism, sport tourism, health/wellness tourism, polar tourism, wilderness tourism and ecotourism, although, as already noted, the distinction between these categories may not always be clear. Importantly, however, they are frequently marketed on the basis of being alternatives to mass tourism. Equally, specific forms of tourism reflect particular interests on the part of tourists – golf tourism, wine tourism or battlefield tourism are just three examples. Malta, for example, started to offer golf tourism in part to compete with surrounding destinations of Cyprus, Sicily, Tunisia, Spain, Portugal and Greece that are also offering the product (Marwick 2000). The main point is, of course, that the demand for tourism is becoming increasingly diverse, providing destinations with opportunities to exploit specific resources or attractions. For example, and as mentioned above, over the past 20 years Cuba has successfully developed resort-based, all-inclusive tourism. However, the country's capital, Havana, has also become an important heritage destination in its own right, with tourism underpinning the redevelopment of the city's Colonial Spanish quarter (Colantonio and Potter 2006) (see also Plate 6.2). Equally, destinations also face the challenges of managing tourism in often fragile environments, as well as remaining distinctive within an increasingly crowded global tourism market.

Plate 6.2 ***Cuba, Havana: Horse and buggy rides for tourists can be contrasted with the local citizens' form of transportation in the background where large trucks have been converted into buses.***

At the same time, traditional forms of tourism are now being consumed in more distant or exotic places. Europeans, for example, now travel to places as diverse as the Caribbean (see Plate 6.3), the Middle East, India, South East Asia, the Indian Ocean or the South Pacific to enjoy sun-beach holidays. Their 'pleasure periphery' has extended well beyond the Mediterranean, potentially bringing the alleged problems of mass tourism to places less able to withstand its impacts. One such destination is Dubai which, though developing a successful tourism sector, faces potential problems in the future (Box 6.2).

Plate 6.3 ***Nassau, Bahamas: Tourists walk through the Prince George Wharf area, which is the main shopping district for tourists arriving by cruise ship.***

Organizing and purchasing tourism

In addition to transformations in *what* tourists are consuming, changes are also occurring in *how* they are organizing and purchasing tourism experiences. Packaged holidays organized and sold by tour operators are still popular; many new or niche tourism products are still sold by tour operators, though mainly smaller, specialist operators are catering for specific markets. However, as a result of widespread Internet use, there is an increasing trend towards tourists organizing and purchasing their holidays independently and online (see example in Chapter 3, p. 58). There is also an increasing tendency for people to take more frequent holidays, supplementing or even replacing the two- or three-week holiday in the summer. A particular phenomenon has been the dramatic increase in short-break holidays facilitated by the growth in low-cost airline operations. Although of less relevance to many less developed countries that lie beyond the effective operational

Box 6.2

Tourism development in Dubai

In recent years, Dubai, one of the seven sheikhdoms making up the United Arab Emirates, has rapidly evolved into the principal international tourist destination in the Middle East. In the early 1980s, Dubai was reliant on oil production which, at that time, accounted for two-thirds of GDP. However, dwindling oil stocks led to the development of tourism and, since then, the emirate's tourism sector has grown dramatically. Tourist arrivals have increased from 374,000 in 1982 to 5,420,000 in 2004, the most rapid growth occurring since 2000; tourist receipts have grown commensurately, exceeding US$1 billion in 2003 and surpassing oil revenues. The growth in arrivals and receipts has, in turn, been driven by an equally dramatic increase in the supply of accommodation. By the end of the 1990s there were some 20,000 hotel rooms on offer, a figure that is expected to quadruple over the next ten years. Flagship projects include the development of The Palm, two man-made islands which will increase Dubai's coastline by 120km, and The World (a collection of islands shaped like countries), both of which will support accommodation and leisure services, while the recently opened Ski Dubai centre (a 400m indoor ski slope) and the planned US$5 billion Dubailand theme park will be among Dubai's major attractions. The majority of visitors are from within the region, although it is becoming an increasingly popular (though expensive) winter-sun destination for European tourists. Average length of stay, however, is relatively short, reflecting Dubai's status as a major air travel hub.

Despite its success in developing tourism, Dubai faces a number of challenges, not the least of which is maintaining sufficient arrivals to fill its burgeoning supply of accommodation. The development of new attractions will be necessary which, along with further accommodation development, will have a major impact on the environment. Moreover, the more it develops, the more it will lose its Middle Eastern 'flavour', while the potential for conflicts between Western tourists and local society and culture may intensify.

Source: Henderson (2006).

range of low-cost flights, such short-break tourism has brought significant economic benefits, for example, to many transitional economies in the former Eastern Europe.

Tourism markets – tourism products

As the supply of tourism products and experiences has become increasingly diverse, the markets for such products have become less distinct. In other words, whereas it was once possible to suggest with reasonable certainty that particular forms of tourism would be consumed by particular groups of tourists, this is no longer the case. On the one hand, experiences that were once the preserve of the better-off, such as cruise holidays or long-haul trips, are now available to the mass market as the industry has recognized the potential of mass marketing 'exclusive' holiday experiences. Equally, the tourism industry is becoming increasingly responsive to the demands of new markets for their products. For example, adventure tour companies that have traditionally organized overland trips for groups of young 'travellers' now sell adventure holidays for families seeking more active or challenging tourism experiences. On the other hand, tourists themselves are consuming tourism in more unexpected ways – one notable example (as mentioned in Chapter 5) is the fact that, of the half a million or so people in the UK taking so-called 'gap year' holidays each year (i.e. a long study or career break), more than half are over the age of 55 and are referred to by some as 'denture-venturers'!

Overall, then, there has been a distinct change in styles of travel and tourism in recent years, with more individualistic, specialist and active forms of tourism becoming predominant. What remains unclear, however, is the extent to which attitudes towards tourism have also changed. In other words, are tourists still motivated by the needs of escapism and/or ego-enhancement, or are they becoming more responsible, aware or 'green'?

Is there a green tourist?

The development of sustainable forms of tourism has long been justified by the belief that tourists in general are becoming greener or more environmentally aware. In other words, reflecting the trends in demand outlined in the previous section, it is claimed by many that the traditional, mass package tourist is being replaced by a more experienced, aware, quality-conscious and proactive tourist consumer; following a shift in general consumer attitudes, tourists 'want more leisure and not necessarily more income, more environmentally sustainable tourism and recreation and less wasteful consumption' (Mieczkowski 1995: 388). At the same time, since the early 1990s innumerable publications have exhorted tourists to act more responsibly, to become 'good' tourists (e.g. Wood and House 1991), though this has been criticized as the 'moralization' of tourism (Butcher 2002). However, the extent to which tourists are becoming greener remains the subject of intense debate, with some commentators concluding that few tourists' holiday choices and behaviour are actually determined by environmental concerns.

There are two arguments in favour of the emergence of the green tourist. First, the rapid growth in demand over the past decade for activities or types of holidays that may collectively be referred to as 'ecotourism' is often cited as evidence that ever-increasing numbers of tourists, as a result of their heightened environmental awareness or concern, are seeking out more appropriate and, in a developmental sense, beneficial forms of tourism. For example, Cater (1993) reported that the number of arrivals at three 'selected ecotourism destinations', namely Belize, Kenya and the Maldives, virtually doubled over a ten-year period from 1981. Similarly, some suggest that participation in ecotourism has increased annually by between 20 to 50 per cent since the early 1980s and now accounts for up to 20 per cent of all international tourism arrivals (Fennell 1999: 163), although others offer more conservative estimates.

Second, surveys consistently point to the alleged emergence of the 'green consumer', both generally and in the more specific context of tourism. In the early 1990s, for example, it was found that increasing numbers of people in the UK considered themselves to be either 'dark green' (i.e. 'always or as far as possible buy environmentally friendly products') or 'pale green' (i.e. 'buy if I see them') consumers (Mintel 1994). A more recent survey by the same organization (Mintel

2007) found that green consumerism is increasing, although consumers are more likely to buy environmentally friendly products to feel good about themselves rather than for altruistic reasons. Similarly, Cowe and Williams (2000) found that one-third of consumers are seriously concerned about ethical issues when shopping. In the tourism context, research has revealed that 64 per cent of UK tourists believe that tourism causes some degree of damage to the environment and that generally UK consumers would be willing to pay more for an environmentally appropriate tourism product (Diamantis 1999). A survey by the charity Tearfund (2000) came to similar conclusions: specifically, it found that 59 per cent of respondents would be happy to pay more for their holidays if the extra money were to contribute to better local wages, environmental conservation and so on. Moreover, it has been claimed that 80 per cent of British tourists would be more likely to book holidays with 'responsible' tourism operators (www.responsibletravel.com).

However, in practice there is little evidence to support either of these arguments. While there is no doubt that the demand for ecotourism and other forms of environmentally aware travel is on the increase, research has consistently failed to demonstrate that tourists who consume such experiences are motivated or influenced by environmental values (Sharpley 2006b). More specifically, studies into the motivation of ecotourists show that the majority seek wilderness scenery, undisturbed nature and the activities that such locations offer as the prime reasons for participating in ecotourism. In other words, it is the pull of particular destinations or holidays (and the anticipated enjoyment of such holidays) that determines participation rather than the influence of environmental values over the consumption of tourism in general. In fact, research into the behaviour of ecotourists in Belize demonstrates that many are motivated by factors other than environmental concern (Box 6.3).

More importantly, perhaps, the assumptions about widespread adherence to the principles of green consumption in general, and the consequential adoption of green tourism consumption practices in particular, also lack foundation. Although there is widespread stated support for green consumerism, it is evident that few consumers regularly buy/consume according to environmental or ethical values. In fact, research has shown that, in the UK, fewer than 1 per cent of people practise green consumerism consistently. Moreover, such behaviour is unlikely to remain constant over time or be applied to all forms of consumption. In short, consumers address environmental issues in complex and ambivalent ways and, as a result, their consumer behaviour is frequently contradictory. That is, their environmental attitudes towards the consumption of different products are likely to vary. Indeed, despite the surveys identifying strong environmental credentials on the part of tourists, UK tourists' spending with responsible tour operators amounted to just £112 million in 2004 (Cooperative Bank 2005), compared with some £26 billion spent on all overseas travel. Thus, although the 'green tourist' undoubtedly exists, the great majority of tourists are more likely not to be influenced by environmental values in their tourism consumption choices and behaviour.

Box 6.3

Ecotourists in Belize

The central American country of Belize was one of the first countries to recognize and embrace the concept of ecotourism as a means of achieving sustainable development through tourism. In the late 1980s, the country adopted an Integrated Tourism Policy to develop ecotourism and, since then, it has become an increasingly popular destination for ecotourists, drawn by the country's natural and cultural (Mayan) heritage and, in particular, the opportunity to explore its famous coral reefs. The subsequent development of tourism in Belize has been widely criticized, specifically the extent to which both national élites and overseas organizations have exploited the country's natural resources, but one study in particular has revealed the actual motivations of some 'eco-tourists' in Belize.

Contrary to the assumption that ever-increasing numbers of tourists are motivated by environmental concerns, research has shown that many visitors to Belize demonstrate behaviour that is stereotypical of the mass tourist. Thus, many tourists are attracted to the country by the opportunity to dive on coral reefs or other natural or cultural experiences, though their visits are framed by the desire for a generic 'Caribbean' experience (beaches, palm trees, reefs, brightly coloured fish), to collect certain sights and experiences, rather than to learn about, experience and contribute to Belize in particular. Few have genuine environmental knowledge and awareness – their stated desire to avoid damaging the coral reefs, for example, is to maintain them as tourist attractions rather than for their intrinsic value – and many act in an inappropriate fashion. Moreover, for many tourists the attraction of Belize lies in the opportunities for hedonistic behaviour, including heavy drinking, drug taking and sexual encounters, either with locals or with other tourists. In short, they lack the environmental awareness and self-reflexivity that might be expected of an ecotourist.

Source: Duffy (2002).

Influencing tourist behaviour: a destination perspective

The form which tourism development in a destination takes is directly related to the nature of tourism consumption in terms of both products demanded and actual behaviour 'on site'. In other words, for the achievement of sustainable tourism development (as defined in Chapter 2), it is necessary not only that tourism should be developed in an appropriate way with regard to character, scale, degree of integration with the local economy and so on, but also that tourists themselves should behave in a responsible manner. As the example of Belize (Box 6.3) demonstrates, developing an ecotourism product does not guarantee sustainable tourism development.

As this chapter has suggested, however, the extent to which tourists are becoming more environmentally aware is unclear. That is, an understanding of the consumption of tourism suggests that responsible tourist behaviour, or purposefully consuming tourism experiences that benefit destination environments and societies, is unlikely to be a widespread phenomenon. How, then, can tourist behaviour be influenced in order to enhance the sustainability of tourism development? Four courses of action deserve attention here:

- Codes of conduct.
- Effective destination planning (according to broader development goals).
- Destination marketing.
- Role of tour operators.

Codes of conduct

In recent years, codes of conduct have emerged as a popular and widely used visitor management tool. This is not to say that they are a new phenomenon – in the UK, for example, the Country Code – advising visitors to the countryside how to act appropriately – was first published in 1953. However, codes of conduct are now considered to be a vital means of promoting responsible behaviour on the part of tourists, and nowadays numerous such codes exist. Some relate to tourism in general, such as an early 'Code of Ethics for Tourists' (O'Grady 1980) which focuses on potential socio-cultural and economic impacts (Figure 6.3), while others may be destination-specific (for example, the Himalayan Tourist Code) or activity-specific; most are directed towards tourists themselves, while some are targeted at the tourism industry or local communities (see also Fennell 2006). The majority of codes are produced by non-governmental organizations or special interest groups; conversely, the tourism industry and governmental bodies have produced few codes (Mason and Mowforth 1996). Collectively, however, there is an enormous and potentially confusing diversity of codes of conduct in tourism. For example,

1. Travel in a spirit of humility and with a genuine desire to learn more about the people of your host country.
2. Be sensitively aware of the feelings of other people, thus preventing what might be offensive behaviour on your part. This applies very much to photography.
3. Cultivate the habit of listening and observing, rather than merely hearing and seeing.
4. Realize that often the people in the country you visit have time concepts and thought patterns different from your own; this does not make them inferior, only different.
5. Instead of looking for that 'beach paradise', discover the enrichment of seeing a different way of life through other eyes.
6. Acquaint yourself with local customs – people will be happy to help you.
7. Instead of the Western practice of knowing all the answers, cultivate the habit of asking questions.
8. Remember that you are only one of the thousands of tourists visiting this country and do not expect special privileges.
9. If you really want your experience to be 'a home away from home', it is foolish to waste money on travelling.
10. When you are shopping, remember that the 'bargain' you obtained was only possible because of the low wages paid to the maker.
11. Do not make promises to people in your host country unless you are certain of carrying them through.
12. Spend time reflecting on your daily experiences in an attempt to deepen your understanding. It has been said that what enriches you may rob and violate others.

Figure 6.3 ***A code of ethics for tourists.***

Source: O'Grady (1980).

one study found that, globally, there are 58 separate codes of conduct relating to whale-watching alone (Garrod and Fennell 2004).

The purpose of codes of conduct is, most usually, to raise awareness among tourists of the need for responsible behaviour. In essence, they are lists of rules for appropriate behaviour in particular contexts, although, as codes, they have no legal status. That is, they depend on voluntary compliance on the part of tourists and, as a result, their effectiveness may be limited, particularly as tourism is seen as a means of escaping from rules and regulation. Therefore, tourists are more likely to respond positively to codes if, in addition to *how*, it is explained *why* they should behave in particular situations.

Destination planning

The purpose of sustainable tourism is to optimize tourism's contribution to a destination's sustainable development. Traditionally, the achievement of this objective has been seen to lie in the development of small-scale, appropriately designed, locally controlled projects, the implication being that these types of tourism development will attract the 'right' kind of tourists. However, it is now accepted that the challenge is to make all forms of tourism sustainable. In other words, different types of tourism development may be more or less suited to the particular resources and developmental needs of different destinations, and therefore, destination planners may be able to influence the nature of tourism consumption through effective planning and promotion of their tourism product.

Tourism in Bhutan, considered in some detail in Chapter 2, is an extreme example of the effective manipulation of tourism consumption by the destination. There, tightly controlled pricing and distribution of holidays, along with strict regulations on tourism activities, ensures that tourists behave in a way that is appropriate to local conditions and needs. Elsewhere, the development of all-inclusive tourism has made a significant contribution to local development while restricting the negative consequences of tourism through what is, in effect, the zoning of tourism development. Importantly, such developments also meet tourists' needs. When the tourism authorities in The Gambia attempted to influence tourism consumption by banning all-inclusive holidays, thereby hoping to encourage expenditure beyond the confines of hotels, demand for holidays in The Gambia declined dramatically.

Equally, there are, of course, numerous examples of destinations or tourism developments that, at the local scale, successfully meet the more typical principles of sustainable tourism (Buckley 2003). The point is that tourism consumption cannot be divorced from local developmental needs, and thus destination planners can play an important role in ensuring that tourism consumption (even if it is mass tourism) is appropriate to those needs.

Destination marketing

As observed earlier in this chapter, a significant influence on the tourism demand process (and the subsequent behaviour and expectations of tourists) are so-called 'effectors' of demand; that is, destination choice is, to a great extent, influenced by the knowledge, images and perceptions tourists have of particular places or experiences. The impact of these images and perceptions on tourist buyer behaviour has been explored widely in the literature (e.g. Pike 2002); however, it is generally accepted that not only are destinations in a position to augment or adapt tourists' images and perceptions, they also have the opportunity to influence the nature of demand for their product. As Buhalis (2000) notes, 'destination marketing facilitates the achievement of tourism policy, which should be coordinated with the regional development strategic plan'.

Despite this potentially influential role of destination marketing in contributing to the appropriate development and consumption of destination experiences, there has been a tendency, particularly in less developed countries, to market or promote places in order to verify, rather than adapt, tourists' images and perceptions (Silver 1993). For example, it has been suggested that much tourism destination marketing in developing countries enhances stereotypical (though not necessarily accurate) images of traditional, undeveloped cultures, and colonial power relations (Echtner and Prasad 2003). As a consequence, tourists may be attracted by marketed representations of the destination that reflect their preconceptions rather than the reality of the place, influencing both their experience and their behaviour.

Nevertheless, destination marketing can play a more positive and proactive role in sustainable destination development. In general, marketing should be resource- rather than demand-based; that is, it should focus on promoting particular attributes of the destination rather than on meeting tourists' perceptions. More specifically, a variety of techniques may be employed, including (see Buhalis 2000):

- *Enhancing and differentiating the product.* Tourists' needs for unique or authentic experiences, for which they may be willing to pay a premium, will be met by a quality, differentiated experience.
- *Appropriate pricing policies.* Pricing is an effective tool in market segmentation (see below) and in ensuring that the full cost to the destination is reflected in the price tourists pay.
- *Embracing the needs of all stakeholders.* The needs of destination-based stakeholders should, in particular, be taken into account in destination marketing.
- *Developing effective public–private sector partnerships.* Such partnerships may facilitate the meeting of broader developmental objectives in the destination.
- *Effective segmentation to target appropriate markets (Leisen 2001).* A closer matching of tourists needs/expectations with destinational resources reduces the potential for negative consequences.

Role of tour operators

Tour operators play a central role in tourism development. They have been described as the 'gatekeepers' of the tourism industry, being able to influence the scale and scope of tourism development as well as the volume and direction of tourist flows. As a result, tour operators are often seen (somewhat inaccurately) as epitomizing mass tourism development, providing cheap holidays to mass markets with little regard for the impacts on destination environments and societies.

In the context of this chapter there is an evident link between tour operators and tourism consumption. Not only do they create and satisfy the demand for tourism experiences, from mass sun-sea-sand holidays to more specialized, niche products, but through the promotion and pricing of their products tour operators are able to dictate the types of tourists travelling to any particular destination. Thus, the success of so-called 18–30s holidays (and the much criticized behaviour of young tourists on such holidays) is very much a result of the manner in which they are promoted.

Increasingly, tour operators are adopting a more responsible approach to their activities, a notable example being the Tour Operators' Initiative, a network of over 20 operators supported by the United Nations Environment Programme (UNEP) that is committed to the development and promotion of sustainable practices in the way they do business. According to a recent publication (UNEP 2005), tour operators should focus on five areas in improving the sustainability of their operations, one of which (customer relations) is concerned with influencing tourist behaviour. Under this heading, a number of principles are proposed:

- *Raise clients' awareness of sustainability issues.* Positive messages in brochures and other information as to how tourists can contribute to the social and environmental well-being of destinations.
- *Develop or adopt a responsible tourism code of conduct.* Tour operators may adopt an existing code or develop their own, communicated to customers in a pre-departure information pack.
- *Communicate sustainability messages throughout the holiday cycle.* Information on sustainability can be provided at all stages of the holiday process, from booking through departure and travel (in-flight messages) to destination experience and post-holiday questionnaires.
- *Reinforce the message between holidays.* Regular mailings to existing customers, such as in-house magazines or promotional leaflets, provide a medium for reinforcing sustainability messages.

The success of any of these methods depends, of course, on tourists' responsiveness to these messages; this, in turn, may depend upon the extent to which a tourist holds strong environmental values. As considered earlier in this chapter, the tourism consumption experience may be dominated by other, more egocentric values, in which case tour operators may be able to contribute to destination sustainability through more direct measures, such as influencing planning, working with local stakeholders or supporting local voluntary/charitable organizations.

Domestic tourism

Not surprisingly, perhaps, most attention in the context of tourism consumption and sustainable tourism development is focused on international tourism, particularly with respect to tourism in the less developed world. On the one hand, international tourism represents, for many countries, a significant export industry and a vital source of foreign exchange earnings. On the other hand, the potential for negative consequences is seen to be greater, and therefore the need for effective planning and management is more pressing.

Frequently overlooked, however, is the contribution that domestic tourism – that is, people consuming tourism in their own countries – can make to sustainable development (Ghimire 2001). In other words, domestic tourism provides many of the benefits of international tourism, such as employment, income, new business development and economic diversification; at the same time it is likely to benefit locally owned and controlled businesses. Therefore, it is important to consider briefly the role of domestic tourism consumption in sustainable development.

Domestic tourism occurs in most, if not all, countries. It is also, in terms of total numbers of trips, significantly greater than international tourism activity; it is estimated that, globally, domestic tourism annually accounts for between six and ten times more trips than international tourism. The reasons for this are self-evident. In wealthier, developed countries, residents frequently make a number of domestic tourist trips each year in addition to overseas holidays, and many only take domestic holidays. For example, research has shown that in the United States, residents collectively make 990 million domestic trips each year compared with some 60 million overseas trips (Bigano *et al.* 2004). In less developed countries, domestic tourism is, for financial reasons, the only option for most residents, although in some countries the number of outbound trips is on the increase. More importantly, in most less developed countries, 'the number of nationals traveling for leisure is considerably higher than the number of international tourist arrivals' (Ghimire 2001: 2).The number of domestic trips is, of course, directly related to population size; according to Bigano *et al.* (2004), China (644 million trips), India (320 million trips), Brazil (176 million trips) and Indonesia (107 million trips) are the four largest domestic tourism markets in the less developed world. Forecasts predict that by 2010 there will be 1.8 billion domestic tourists in China as more and more Chinese make repeat visits within their own country (Elegant 2006). The growth in Chinese domestic tourism is directly related to a dramatic increase in automobile sales and highway expansion. China's 21,100 miles of highway are predicted to more than double by 2020 (Elegant 2006). However, even in smaller countries such as Turkey, domestic tourism is a growing sector and a potential contributor to regional development (Seckelman 2002).

There are a number of potential benefits of promoting domestic tourism consumption:

- It can act as a catalyst of local or regional economic growth, particularly in peripheral rural areas; this has been a particular driver of domestic tourism in China, for example.

- It may spread wealth from richer, urban areas as emerging middle classes engage in domestic leisure travel.
- In some instances, the potential socio-cultural impacts may be fewer than with international tourism development, while opportunities for community-based tourism may be greater.
- Domestic tourism may provide the basis for developing greater social cohesion, national cultural identity or, as in the case of South Africa, encouraging national reconciliation (Koch and Massyn 2001).

At the same time, of course, domestic tourism development faces many similar challenges to international tourism development, including environmental degradation, social pressures and cultural commodification, while local tourism resources (e.g. businesses, land, attractions) may come to be owned or controlled by urban élites. The key to success lies in effective planning and management, as has been adopted in countries such as Brazil and Mexico; generally, however, there is still a lack of knowledge and understanding about the benefits and costs of domestic tourism in the less developed world.

Tourism consumption has changed over time and, as indicated in Chapter 4, tourists have a role to play in the development process. What products they purchase (see Plate 6.4), how much they spend, how they interact with the host community and how they behave all have the potential to influence developmental goals in the destination. The following chapter will expand on these issues by exploring the broader impacts of tourism.

Plate 6.4 ***South Africa, near Pretoria: Shops selling souvenirs to tourists. Note the taxidermy animals for sale.***

Discussion questions

1. **What are the emerging trends in tourism demand?**
2. **Are tourists really interested in becoming green?**
3. **What are the advantages and disadvantages for targeting a specific socio-demographic tourist segment or tourist type such as cultural tourists or adventure tourists?**
4. **How can marketing be used to control tourist behaviour?**

Further reading

Butcher, J. (2003) *The Moralisation of Tourism. Sun, Sand . . . and Saving the World?* London: Routledge. [This book sets out to challenge the view that not only are tourists becoming 'greener', but also that tourists and the tourism industry should adopt a more responsible approach to the consumption and development of tourism. It provides a fascinating yet controversial critique of contemporary, responsible approaches to tourism.]

Mann, M. (2000) *The Community Tourism Guide*, London: Earthscan. [Essentially a guide to so-called 'real' holidays, or responsible/sustainable community-based tourism products, this book sets out the reasons why tourists should behave responsibly and provides numerous examples of responsible tourism in practice.]

Sharpley, R. (2003) *Tourism, Tourists and Society* (3rd edn), Huntingdon: Elm Publications. [This book explores the relationship between tourists and society from two perspectives: the influence on society of tourists/tourist behaviour, and the influence (impacts) of tourists on society. It provides an in-depth analysis of tourism demand and behaviour, and particular attention is paid to the nature of, and influences on, tourism consumption within a postmodern context.]

Swarbooke, J. and Horner, S. (1999) *Consumer Behaviour in Tourism*, Oxford: Butterworth-Heinemann. [This book provides a thorough and broad introduction to tourism consumption. Introducing the main theories and concepts relevant to the consumption of tourism, it considers principal contemporary issues and debates, and refers to a wide range of literature.]

Websites

The website of The International Ecotourism Society (TIES), a global network of individuals, institutions and tourism industry representatives that promotes and supports responsible practices in tourism: www.ecotourism.org.

ECPAT is the leading children's rights organization that seeks to prevent the exploitation of children, including through tourism: www.ecpat.org.uk.

Partners in Responsible Tourism (PIRT) is an example of a local network that promotes culturally and environmentally responsible tourism: www.pirt.org.

Responsible Travel is an online travel agency specializing in 'responsible' travel and holidays: www.responsibletravel.com.

Tourism Concern is a UK-based group that seeks to raise awareness of, and solutions to, the negative social, cultural, environmental and economic consequences of tourism: www.tourismconcern.org.uk.

7 Assessing the impacts of tourism

Learning objectives

When you have finished reading this chapter, you should be able to:

- **Appreciate the wider social, political and economic contexts within which tourism's impacts occur;**
- **Identify the main socio-cultural, environmental and economic impacts of tourism;**
- **Understand how local residents react to the impacts of tourism;**
- **Be aware of measures to help minimize the negative impacts of tourism.**

The central theme of this book is that tourist destinations, particularly those in the developing world, face a dilemma. On the one hand, tourism is widely seen as an effective means of achieving development; it represents a potentially valuable source of income and employment and a driver of broader economic, infrastructural and socio-cultural development. On the other hand, such development cannot be achieved without cost. Not only does tourism, as a resource-based industry, inevitably exploit or 'use up' resources, be they natural, man-made or human (McKercher 1993), but also the activities of tourists themselves may have significant impacts on the destination, on its environment and local communities. If not controlled or managed, these costs may outweigh the benefits of tourism development and, in the longer term, reduce the attraction of the destination to tourists.

The dilemma facing destinations, therefore, is how to meet the broader developmental objectives of each destination by optimizing the contribution of tourism while, at the same time, keeping the costs or negative consequences of tourism development to a minimum. In other words, the challenge for destinations is to achieve sustainability of the tourism sector, fundamental to which is the need to balance the positive and negative impacts of tourism within the context of broader development goals.

The overall purpose of this chapter is to consider the nature of these impacts and to explore the ways in which they may be planned and managed effectively within a sustainable developmental context. First, though, it is important to emphasize that the impacts of tourism cannot, or should not, be viewed in isolation. In other words, the impacts of tourism or, as Wall and Mathieson (2006) suggest, the 'consequences' of tourism – the term 'impacts' is often seen as having negative connotations – are frequently considered generically in the tourism literature, with typical impacts simply listed and described. However, the way in which such impacts are perceived, the extent to which they are felt, and the manner in which they are responded to varies enormously from one destination to another. In other words, tourism impacts arise from a complex interrelationship between the destination, the tourism industry and tourists, and should be viewed within the broader and dynamic economic, social and political contexts within which they occur. The overall outcome of the impacts will influence the contribution of tourism to development. Therefore, the first section of this chapter provides a framework for assessing the impacts of tourism.

Tourism impacts: a framework

The impacts of tourism are felt across the tourism system; moreover, such impacts may be both positive and negative. In tourism-generating regions, for example, outbound tourism is an important source of employment in travel retailers, tour operators, airports, transport operators and so on. Conversely, air transport – the 'transit region' (Leiper 1979) – is seen as a major contributor of greenhouses gases and, hence, a significant negative impact. Nevertheless, the impacts of tourism are usually considered in the context of the destination where tourism development occurs, where tourists come into contact with local people and the environment, and where there is arguably the greatest need to identify, measure and manage such impacts, both positive and negative.

In assessing the impacts of tourism, a useful starting point is to consider the destination as an overall tourism environment. Frequently, the 'environment' is thought of simply in terms of the physical attributes (natural and built) of the destination. However, tourists seek out attractive, distinctive or authentic tourism environments or destinations, which, as Holden (2000: 24) observes, possess 'social, cultural, economic and political dimensions, besides a physical one'. In other words, the tourism environment may be defined as:

> that vast array of factors which represent external (dis)-economies of a tourism resort: natural . . . anthropological, economic, social, cultural, historical, architectural and infrastructural factors which represent a habitat onto which tourism activities are grafted and which is thereby exploited and changed by the exercise of tourism business.
>
> (EC 1993: 4)

Thus, from a tourist's perspective, the social or cultural aspects of a destination are usually inseparable from its physical aspects in terms of their overall experience of that destination. The important point, however, is that *perceptions* of the tourism environment may vary significantly; that is, different groups will perceive or value the tourism environment in different ways. As we saw in Chapter 6, tourists' attitudes towards the destination may vary enormously, as might their subsequent behaviour while on holiday. Equally, there is likely to be a distinction between the ways in which local communities and tourists perceive the destination environment, reflecting local cultural/environmental values, economic need and so on as illustrated in Chapter 5. For example, while tourists may value a pristine or undeveloped environment, locals may view it as a legitimate resource for exploitation as they seek to enhance their lifestyle or achieve social and economic development. Conversely, local communities may strongly value or hold sacred particular places or environments, and seek to protect them from tourism exploitation or development. Uluru (Ayers Rock) in Australia is a well-known example of the latter (Brown 1999), while the authorities in Bhutan have long sought to protect sites of cultural or environmental significance from tourism exploitation (Dorji 2001). At the same time, it has been found that grass-roots activist groups in southern Europe, specifically in Greece, Spain and Portugal, have increasingly engaged in demonstrations and other forms of protest against environmentally damaging tourism developments (Kousis 2000).

In other words, when assessing the impacts of tourism, it is essential to consider the broader social, political and economic context of the destination and to recognize that planning and management decisions with respect to tourism development should reflect that local context rather than external, often Western-centric values. This is, of course, particularly the case in less developed countries, where the need for socio-economic development and modernization may be considered to be more important than providing tourists with traditional or authentic experiences; the example of Singapore where, in the 1970s, traditional Chinese shops and markets were replaced with contemporary buildings and experiences (Lea 1988) has since been repeated in many other places.

It is also important to consider the impacts of tourism within a wider national and global context. Tourism destinations do not exist in isolation; they are part of an international tourism system within which major multinational corporations frequently play a dominant role, resulting in a condition of dependency in many destinations (see Chapter 1). At the same time, the international tourism system is part of a global political, economic and socio-cultural system. Consequently, two points should be noted:

- The impacts of tourism in a particular destination may often be influenced by factors beyond that destination's control. For example, the USA's continuing embargo on Americans travelling to Cuba has undoubtedly affected the development of the island's tourism sector; similarly, campaigns by a number of international organizations and pressure groups to boycott tourism to Myanmar have also had some influence on the decline in international tourist arrivals in that country since the late 1990s (Henderson 2003).
- Impacts that are attributed to tourism are often the result of wider, global

influences (see Chapter 3). For example, and as we shall see shortly, international tourism is often blamed for weakening local culture, threatening traditional social structures or for introducing Western cultural practices to local communities. However, the alleged globalization of Western culture (Held 2000), underpinned by dramatic advances in information and communication technologies, is seen as a more powerful influence on socio-cultural change in many less developed countries.

Figure 7.1 provides a simple framework for assessing the impacts of tourism at the level of the destination. This illustrates that the nature of tourism impacts, the

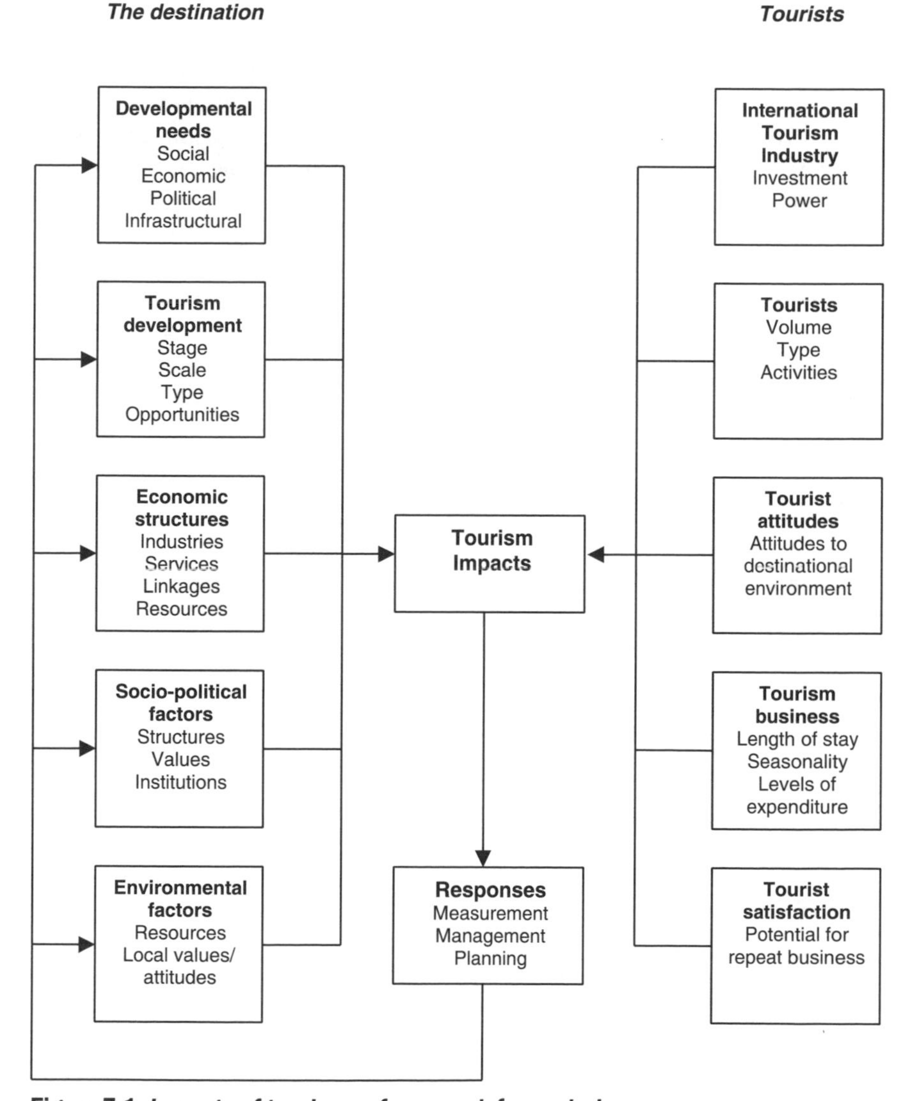

Figure 7.1 ***Impacts of tourism: a framework for analysis.***

manner in which they are perceived, and subsequent planning and managerial responses are a function of both destinational and tourism developmental factors. Implicit in this model is the influence, referred to above, of local needs and values with regard to perceptions of and responses to the impacts of tourism, although the extent to which such influence may be exercised will reflect local political and social structures. However, Figure 7.1 also illustrates the tangible factors that may determine the nature and extent of tourism's impacts in destination areas. These fall under two broad headings, namely tourists, and destination factors and characteristics, and are summarized below.

Tourists

There is not necessarily a causal effect between the volume of tourists and degree of impact; in some instances, such as culturally or environmentally fragile areas, a very small number of tourists may have a major negative impact, whereas in other areas, large numbers of tourists may cause relatively limited negative impacts yet provide significant economic/developmental benefits. Nevertheless, greater negative impacts are usually associated with high-volume tourism, particularly when such tourism is associated with supply dominated by international tourism businesses, and hence a condition of dependency (see Chapter 1). More certain is the fact that, as discussed in detail in the previous chapter, the types of tourists, their attitudes and their subsequent activities will have a direct influence on the nature and extent of destinational impact. Figure 7.2 summarizes the relationship between tourists' environmental values and their subsequent actions or impacts.

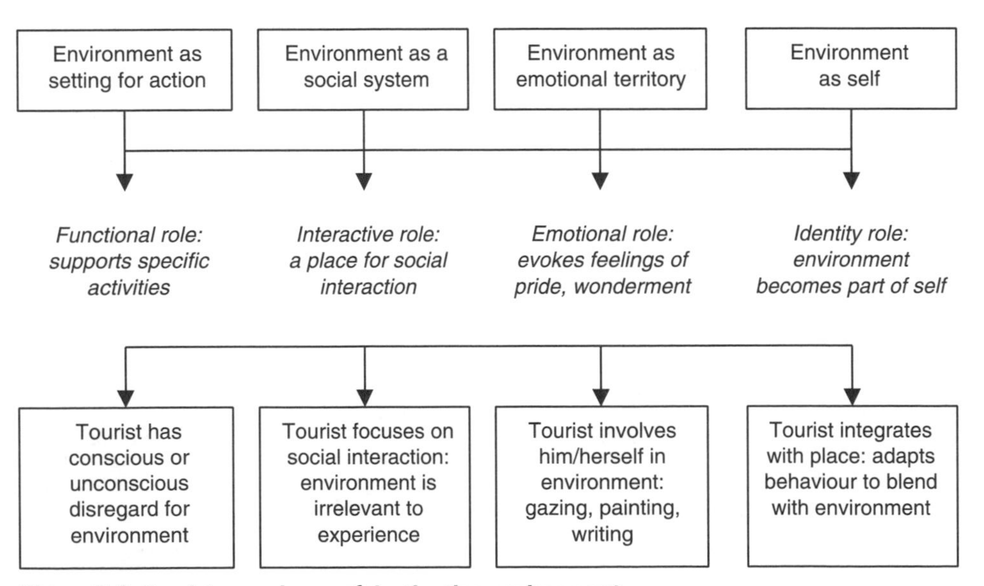

Figure 7.2 ***Tourist experience of destination environment.***

Source: After Holden (2000: 49–50)

Destination factors and characteristics

There are a variety of destination characteristics that determine both the positive and negative impacts of tourism. These are considered elsewhere (Burns and Holden 1995; Wall and Mathieson 2006), but include the following:

- *Character and sensitivity of the environment.* Some environments or ecosystems are more fragile, less robust or more sensitive to change than others, or may take longer to recover from physical damage.
- *Economic structures and stage of development.* The diversity of the local economy, the availability of investment funding, the import–export balance and the ability to meet the tourism sector's requirements, as well as the overall extent of economic development, will determine the degree of economic benefit enjoyed.
- *Political structures and processes.* Tourism development is influenced by local and national political structures, the extent of political engagement or influence in tourism planning and management, and the nature of policies directly and indirectly affecting tourism.
- *The nature, scale and rate of growth of tourism development.* An evident link exists between the character and scale of tourism development and the subsequent impacts on the destination. However, if tourism development outstrips appropriate infrastructural development, such as sewage systems, negative impacts may be enhanced.
- *Social structures.* The size and structure of local communities, their cultural practices, moral codes, religious affiliations, language and so on (particularly in comparison to those of tourists) are fundamental to the degree of 'felt' impact at the destination.

There are, then, numerous influences on the extent to which tourism impacts both positively and negatively on a destination, and the ways in which those impacts are perceived and responded to by local communities and visitors alike. The next task is actually to identify the 'typical' impacts of tourism.

The impacts of tourism: an overview

A detailed discussion of the impacts of tourism is well beyond the scope of this chapter and readers should refer to more detailed reviews of the topic in the literature. Many general tourism textbooks cover the topic more than adequately (see e.g. Holloway 2002; Cooper *et al.* 2005; Page and Connell 2006), while innumerable books and articles address it specifically (in particular, Wall and Mathieson 2006). Nevertheless, the relationship between tourism and development cannot be fully understood without an appreciation of the principal impacts of tourism, both positive and negative. Indeed, if 'development', as defined at the beginning of this book, is the intended overall outcome or impact of tourism; the challenge is to manage the more specific impacts effectively so that tourism's contribution to development is optimized. The purpose of this section, therefore, is to introduce the main impacts of tourism.

Typically, tourism's impacts are assessed under three broad headings, namely economic impacts, physical (environmental) impacts and socio-cultural impacts. For convenience, this section does likewise.

Economic impacts

The driving force behind the development of tourism is its potential contribution to destination economies. This is particularly so in less developed countries, where tourism is seen as an effective (and sometimes the only) catalyst of economic growth and wider socio-economic development. For many such countries throughout the developing world, tourism represents an economic lifeline, its contribution usually measured in terms of earnings from tourism (tourism receipts), export earnings (balance of payments), contribution to GNP and employment generation. Many countries have adopted the Tourism Satellite Account system endorsed by the World Tourism Organization for tracking tourism-related statistics. In some cases, such as Mexico, Thailand, the Seychelles and Fiji, tourism has proved to be an economic success, both as a specific sector and as a driver of development; elsewhere, however, it has failed to achieve its developmental potential, despite providing an important source of foreign exchange earnings and employment. As noted in Chapter 2, The Gambia has developed little over the past decade, despite the relative importance of tourism to its economy, while Turkey is another example of a country that has enjoyed limited success in exploiting its tourism sector as an engine of national development (Tosun 1999).

In other words, although tourism undoubtedly makes a measurable contribution to destination economies, neither the magnitude of tourism's economic impacts nor its role in stimulating wider socio-economic development can be taken for granted. Not only are there various economic costs associated with tourism development, which, in effect, limit its net economic benefits, but there are also many factors that may reduce its broader developmental contribution (see Figure 7.3). As a result, care should be taken in assessing the economic impacts of tourism (Wall and Mathieson 2006: 79); short-term economic benefits must be measured against economic (and non-economic) costs and the extent of the longer term contribution of tourism to development.

The principal economic impacts of tourism are as follows.

Economic benefits

Contribution to balance of payments/foreign exchange earnings

In 2004, total international tourism receipts amounted to US$623 billion, representing 6 per cent of global exports of goods and services (or 8 per cent when international fare receipts are included), and over 32 per cent of global exports of services alone. Therefore, tourism is a valuable source of foreign exchange earnings, particularly earnings of 'hard' currencies. The development of tourism in Cuba since the early 1990s, for example, has been driven almost entirely by the

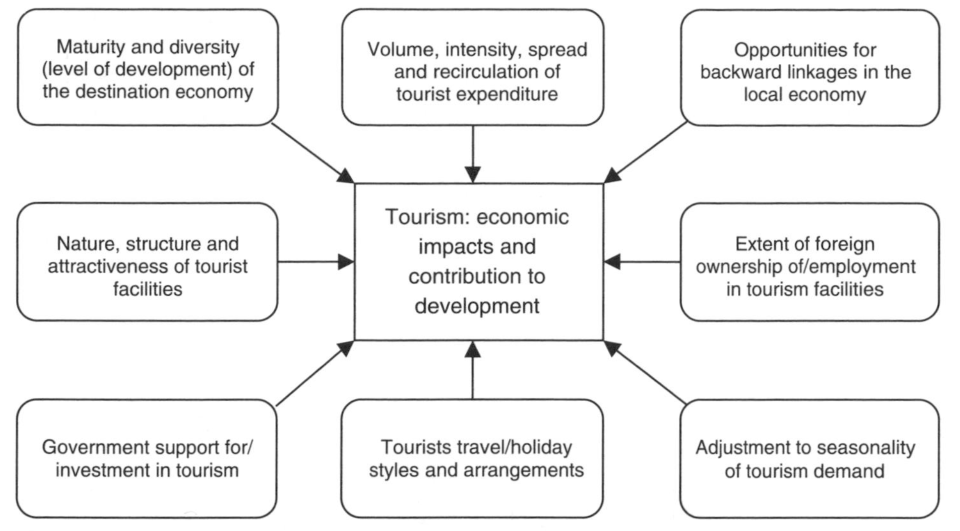

Figure 7.3 ***Factors that influence the economic impacts of tourism.***

Sources: Adapted from Lea (1988); Wall and Mathieson (2006: 90).

country's need for hard currency. However, three points should be noted. First, developing countries benefit collectively from less than 40 per cent of total international receipts, reflecting the inequitable redistribution of (Western) wealth through tourism. Second, a country's international tourism receipts must be measured against its own expenditure on overseas travel. For many developing countries, the 'travel balance' shows a healthy surplus, although, as a country becomes wealthier and its citizens travel overseas more frequently, this surplus may become smaller. Finally, and as explained below, it is *net* contribution to the balance of payments that must be considered – many countries have significant import costs to meet the needs of tourists.

Income generation

Tourism, both international and domestic, is a source of income for businesses and individuals that supply goods and services to tourists. The primary source of such income is direct expenditure by tourists on goods and services, including accommodation, transport, entertainment, food and beverages, and shopping; however, there are also indirect (secondary) and induced (tertiary) effects of tourism spending. Indirect effects relate to the expenditure by tourism businesses on goods and services in the local economy; hotels, for example, purchase food, beverages, equipment, power and water supplies and so on, as well as the services of the construction industry. These suppliers need to purchase goods and services in the local economy, and thus the process of expenditure continues through successive rounds. Eventually, income earned by local people as a result of these rounds of expenditure is spent in the local economy as induced spending. In other words, the original amount spent by tourists is multiplied by a particular amount that reflects the extent of subsequent economic activity which itself is determined by the characteristics of the local economy. This 'multiplier effect' is an important tool in calculating the overall economic benefit of tourism, and identifies the value of the overall contribution of a country's 'tourism economy' as opposed to its tourism receipts. The process is summarized in Figure 7.4.

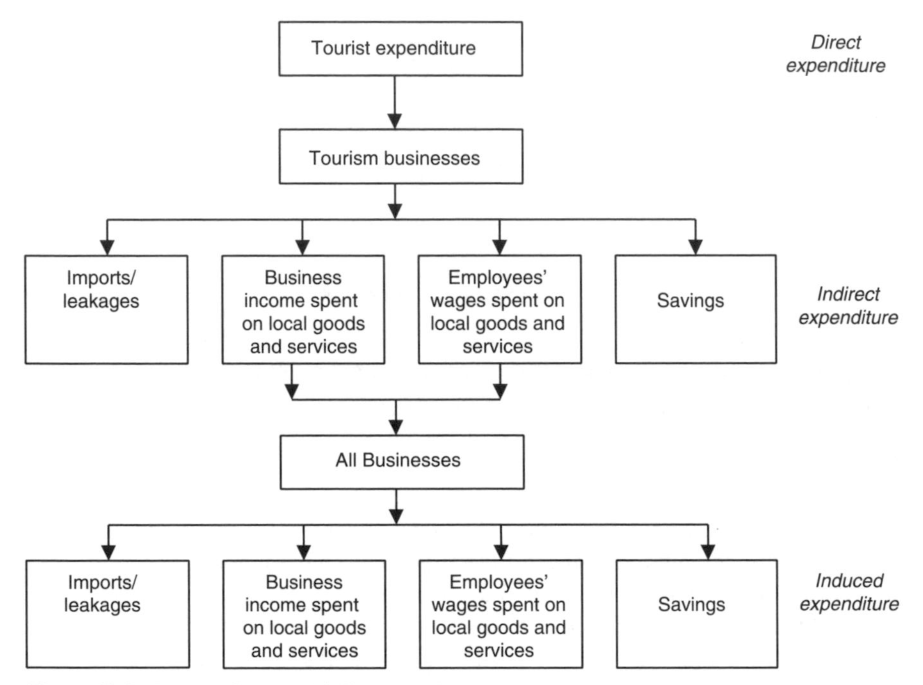

Figure 7.4 ***The tourism multiplier process.***

Source: Adapted from Cooper *et al.* (2005: 165).

Tourism may also be a significant source of revenue for governments. In addition to income tax paid by local tourism workers, for example, sales taxes are often imposed on goods and services sold in tourist establishments or imported goods may be subject to import duties. Frequently, such revenue is used to fund the further development or promotion of tourism. Governments may also charge development fees relating to the construction of tourism developments.

Importantly, data identifying the income generated by tourism (often expressed in terms of contribution to GDP) do not reveal the spread of income around an economy or, in other words, the extent to which income from tourism is shared among the local population. For example, those excluded from the formal tourism economy may be unable to benefit from tourism; hence the potential contribution of pro-poor tourism schemes designed to enhance the earning opportunities of people who are not formally employed within the tourism sector (see e.g. Bah and Goodwin 2003).

Employment generation

Tourism is widely considered to be a labour-intensive industry and thus an effective generator of employment opportunities both in the formal and informal sector. In some developing countries, wages in the tourism industry may be much higher than those in other sectors, thereby attracting workers from other sectors. In Bali, Indonesia, Cukier (2002) found informal tourism employment to be very important and she recommends that this sector should be supported in government policy. The full employment-generation impact of tourism may be revealed by an 'employment multiplier' similar to that described above. However, care must be taken in assessing the contribution of tourism to employment, for a number of reasons:

- The number of jobs created is dependent on the nature and scale of tourism development; some forms of tourism are more labour-intensive than others.
- Jobs in tourism tend to require lower levels of skills and training, while opportunities for promotion may be limited.
- Tourism employment also tends to be characterized by lower paid, casual or part-time jobs. In many destinations it is also highly seasonal. Thus, the contribution to full-time, permanent employment may be more limited than is apparent, particularly as many jobs may be taken by people who do not belong to the official workforce (e.g. students, retired people or 'informal' workers).
- Similarly, tourism may simply attract workers from other, more traditional sectors of the economy, such as agriculture, thereby not only having little impact on unemployment levels but also leading to labour shortages in those sectors.
- Employment in formal tourism may be very insecure with workers not having formal contracts. Tourism Concern is running a campaign to raise the issue of employment conditions in tourism in developing countries (see details on its website).

Entrepreneurial activities/backward linkages

A key contribution of tourism as an agent of development is its potential to stimulate backward linkages or entrepreneurial activity throughout the local economy. This may take the form of informal tour guides or artists selling crafts to tourists as souvenirs (see Plate 7.1). The development of tourism requires a variety of goods and services to both establish the sector and to meet the needs of tourists. Indeed, such linkages are a fundamental element of sustainable tourism development (Telfer and Wall 1996), but are dependent on a number of factors, including:

- The types of goods and services required and the ability of local producers/suppliers to provide them in terms of both quantity and quality.
- The scale and rate of tourism development; rapid, large-scale development tends to outstrip limited local supplies of goods and services.
- The type of tourism in the destination, hence the types of goods and services required.

Plate 7.1 ***Indonesia, Bali: Local entrepreneurs are making dyed textiles to be sold as souvenirs to tourists. They are laid out on the riverbank to be dried after the dying process.***

Economic costs

Although attention tends to be focused on the economic benefits of tourism, it is also important to recognize that the development of tourism is associated with a variety of economic costs, thereby reducing the contribution of tourism to development.

Leakages/propensity to import

Related to tourism's contribution to the balance of payments, an obvious cost is the import of goods and services, or so-called 'leakages', to meet the needs of the tourism sector. Such leakages may be significant in smaller, developing countries with limited local economic sectors, and are greater if tourists travel with overseas tour companies and/or stay in foreign-owned hotels. Torres (2003), for example, found that the state-planned tourism development in Cancún failed to stimulate local agricultural development in Quintanna Roo, resulting in a loss of opportunity for local agriculture and a loss of potential tourism benefits for the local region.

Overdependency on tourism

An inherent danger of developing tourism is that the local economy, perhaps once dependent on a single primary product affected by global commodity prices, becomes overdependent on tourism, and therefore highly susceptible to changes in tourism demand. Such changes reflect the vulnerability of tourism to a variety of influences, such as natural disasters, terrorist activity, global economic downturn or, simply, changes in fashion. Dependence on tourism may be measured by its contribution to GDP which, as Table 7.1 demonstrates, may be significant, particularly in small island states.

Table 7.1 Travel and tourism economy as percentage of total GDP

Rank	Country	% GDP	Rank	Country	% GDP
1	Macau	93.6	11	Guadeloupe	44.1
2	Antigua and Barbuda	85.4	12	Virgin Islands	42.9
3	Aruba	78.0	13	Barbados	41.4
4	Anguilla	74.7	14	Angola	40.5
5	Maldives	66.6	15	Cayman Islands	34.4
6	British Virgin Islands	54.7	16	St. Vincent & Grenadines	33.8
7	Seychelles	54.1	17	Fiji	33.1
8	Saint Lucia	51.0	18	Jamaica	33.1
9	Bahamas	50.1	19	Other Oceana	31.8
10	Vanuatu	47.0	20	Grenada	29.9

Source: WTTC (2006: 46).

In The Gambia, for example, a military coup in 1994 resulted in the collapse of its tourism industry during the 1994 to 1995 season, causing widespread economic problems (Sharpley *et al.* 1996). The more recent coup in Fiji in December 2006 also had an immediate impact on the tourism sector. It was reported, for example, that in the week following the coup hotel occupancy had fallen to 25 per cent and that the country was losing some $1.3 million a day in tourist expenditure (TVNZ 2006). Similarly, the terrorist bombing in Bali in October 2002 had devastating economic consequences for the island's economy which, at that time, depended on tourism for some 50 per cent of its income and 40 per cent of direct employment (Hitchcock and Darma Putra 2005).

Inflation

Tourism development may lead to inflation, particularly with respect to retail prices during the tourist season and to property/land values in popular tourist areas.

Opportunity costs

Opportunity costs refer to the economic benefits of tourism compared with the potential benefits forgone of investing in an alternative economic sector (or opportunity). Little research has been undertaken into opportunity costs, although ignoring them may over-emphasize the assumed economic benefits of tourism.

Externalities

The development of tourism inevitably results in externalities, or incidental costs borne by local communities. These include costs such as additional refuse collections during the tourist season, policing, traffic management and health services.

Physical impacts

It is inevitable that tourism brings about physical or environmental impacts. The development of tourism infrastructure, facilities and attractions transforms natural environments, while the presence of tourists and their various activities have further and continual impacts on both the natural and built environment. It is not surprising, therefore, that not only have the physical impacts of tourism long been recognized, but also a significant degree of attention has been paid to them in the literature (see e.g. Hunter and Green 1995; Mieczkowski 1995; Holden 2000; Wall and Mathieson 2006). Moreover, concern for the physical impacts of tourism lies at the heart of the sustainable tourism development debate – fundamental to the sustainability of tourism is the maintenance and health of its physical resource base.

Typically, the study of tourism's physical impacts focuses on negative impacts, or environmental costs. At the same time, the analysis is frequently structured around particular impacts, such as pollution or erosion, or around the constituent elements of the natural or built environment, often in isolation from the wider political or socio-economic contexts of the destination. However, a number of points deserve mention:

- Tourism development may, in fact, encourage environmental conservation and improvement as well as park creation; that is, tourism may have positive environmental consequences (see Plate 7.2).

Plate 7.2 ***Argentina, Iguazu Falls: A World Heritage Site that lies on the Argentina–Brazil border. Parks protect the jungle ecosystem on both sides of the Falls. Parks can play an important role in resource protection.***

Source: Tom and Hazel Telfer.

- There is often no baseline for measuring or monitoring environmental change that results from tourism development, particularly against local community perceptions of 'acceptable' change or damage.
- Tourism development may have both immediate and secondary environmental impacts; that is, it is sometimes difficult to isolate one type of impact from another.
- Environmental change in destination areas may not, in fact, be caused by tourism development, but by other human or economic activities.

Rather than simply isolating and describing specific types of impact, therefore, it is more useful to adopt an holistic approach to the study of tourism's physical impacts within the overall context of tourism development at the destination. One framework for doing so was developed by the OECD in the late 1970s, and this remains perhaps the most comprehensive and integrated model for assessing tourism's physical impacts (Lea 1988; Pearce 1989). Key to the model is the identification of a number of tourism-generated stressor activities (e.g. resort construction, generation of wastes and tourist activities), the nature of the stresses themselves and the primary (environmental) and secondary (human) responses to this environmental stress. Figure 7.5 presents an adapted version of the OECD framework, embracing the destinational influences discussed earlier in this chapter. The principal physical impacts of tourism may be summarized under the four 'stressor activities' highlighted in Figure 7.5.

Permanent environmental restructuring

The development of tourist destinations requires the construction of facilitics and attractions (hotels, resorts, restaurants) and associated infrastructure, including roads, railways, airport terminals and runways, harbours and marinas and so on. Such development results, of course, in the permanent transformation of the physical fabric of an area as the built environment expands, 'using up' natural ecosytems or taking large areas of land out of primary (agricultural) production. An immediate effect is a change in the visual quality of an area (sometimes referred to as 'architectural pollution') but, in the longer term, significant changes may occur within the natural environment as wildlife habitats are threatened or as ecosystems are damaged by physical construction, or by associated impacts such as pollution. For example, fragile marine environments, including coral reefs, may be seriously damaged by increased sediment loads in coastal waters surrounding tourist developments (Mowl 2002). In response, conservation measures, environmental improvement schemes or visitor management programmes may be needed to prevent further environmental deterioration.

Generation of waste and pollution

Tourism is a significant generator of waste materials, manifested primarily in the pollution of land, air and water resources. By definition, tourism involves transport, and it is not surprising that different modes of tourist transportation collectively represent a major source of air and noise pollution. In recent years,

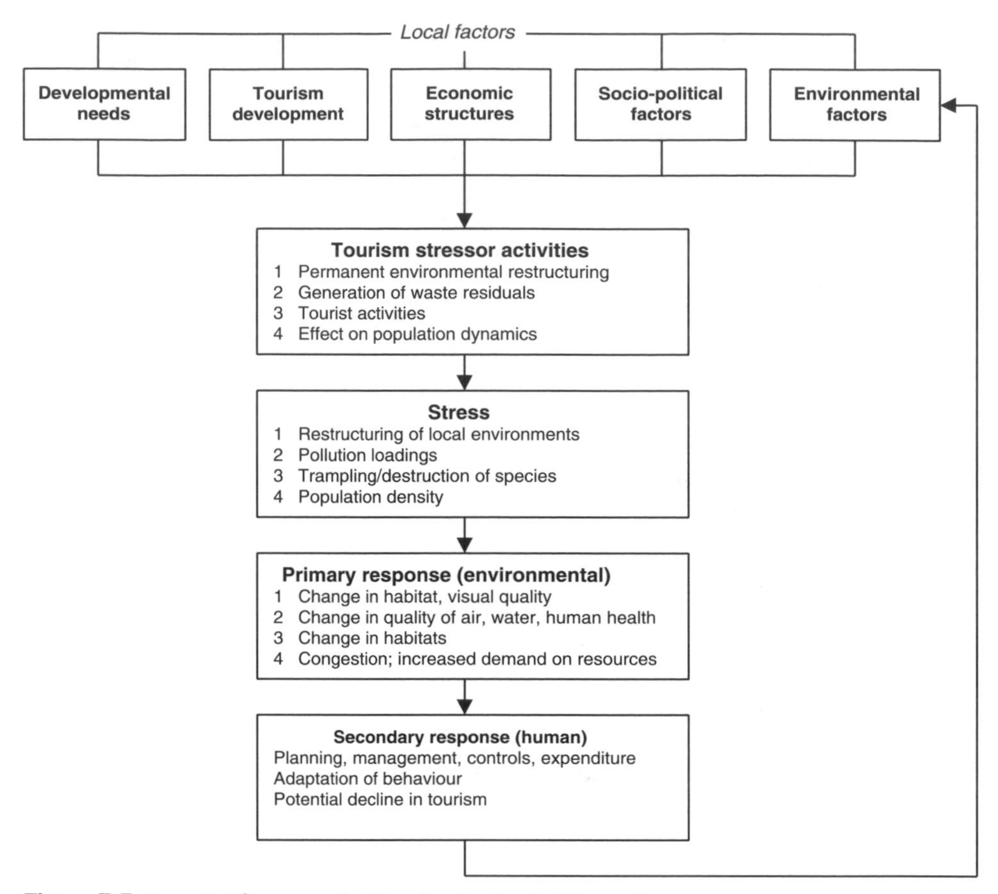

Figure 7.5 ***A model for assessing tourism's physical impacts.***

Source: Adapted from OECD (1981).

increasing attention has been paid in particular to the environmental impact of aviation emissions, air travel being seen by many as the world's fastest growing source of greenhouse gases. Currently, the global commercial jet aircraft fleet annually generates over 700 million tons of carbon dioxide, the major greenhouse gas (www.greenskies.org), although the aviation industry is quick to point out that just 5 per cent of global warming is accounted for by commercial air transport. However, it is estimated that continuing growth in air travel will increase this contribution to global warming to 15 per cent by 2050. In comparison, cars currently produce 10 per cent of global greenhouse gas emissions, though half of global car emissions are produced in the USA alone (Borger 2006). Nevertheless, it is evident that tourism-related transport may make a significant contribution in the longer term to air pollution and, hence, climate change (see the special issue of the *Journal of Sustainable Tourism*, vol. 14, no. 4 (2006) and the edited volume by Hall and Higham 2005). Water pollution also occurs as a result of tourist transport. It has been estimated, for example, that the Caribbean region receives a total of

63,000 port calls annually from cruise ships, generating some 82,000 tons of garbage (Campbell 1999).

Resort and other tourism-related built environments also create waste and pollution. For example, marine and freshwater pollution is a common problem resulting from inadequate sewage treatment facilities in resort areas, meaning that raw or untreated sewage is pumped into lakes or the sea. As Lea (1988) notes, this not only impacts negatively on marine ecosystems, but may also represent a health hazard for tourists swimming in dirty water. Tourism also generates significant amounts of solid wastes, the improper disposal of which may lead to land contamination, unattractive or degraded environments, and health hazards both for humans and wildlife. For example, Holden (2000) observes that elephants in the Masai Mara game park in Kenya have reportedly been poisoned by eating zinc batteries left on a rubbish tip outside a tourist lodge.

Tourist activities

Perhaps the most widespread, and most documented, physical impacts of tourism result from the activities of tourists themselves. Some such impacts are caused either wilfully or through ignorance; for example, tourists walking on or plundering coral reefs is a major problem in many countries. Indeed, it has been reported that, of the 109 countries possessing coral reefs, in 90 of them the reefs have been damaged by cruise ship anchors and, in particular, tourists breaking off chunks of coral (UNEP 2002) while, according to Goudie and Viles (1997), 73 per cent of coral reefs off the coast of Egypt have been adversely affected by tourist activities.

More commonly, the impacts of tourist activities are unintentional, resulting simply from tourists 'being there' in numbers that exceed an area's 'carrying capacity' (see below). Both natural and built environments are adversely affected by the presence of tourists, with the erosion or destruction of fragile areas through walking, trekking, cycling or other activities being a common problem. In some instances, particular activities have a variety of environmental impacts; trekking in remote mountain regions not only results in physical damage to paths and trails, but litter is also a serious problem. For example, some trekking trails in the Peruvian Andes have been nicknamed 'Coca-Cola trails', reflecting the volume of discarded empty drink bottles along the routes. At the same time, as the case of the Nepalese Himalayas demonstrates (Box 7.1), tourism may also lead to significant deforestation problems which, in the longer term, have wider social and environmental consequences.

There are, of course, numerous other environmental consequences arising from the activities of tourists, from forest fires caused by discarded cigarettes or graffiti on historic structures, to the disruption of breeding or feeding patterns of wildlife in safari parks (Shackley 1996). The challenge for most destinations, however, is how to manage or control environmentally damaging tourist activities without discouraging or preventing tourism in the first place.

Box 7.1

Impacts of trekking in the Nepalese Himalaya

The small Himalayan kingdom of Nepal lies land-locked between India to the south and Tibet to the north. It is one of the poorest countries in the world – per capita income is just US$250 – and it faces many of the challenges of underdevelopment. For example, average life expectancy is 60, infant mortality is 61 per 1,000 births and the literacy rate (adults over 15 years of age) is 44 per cent. With few natural resources, the country relies heavily on international aid which, in 2004, amounted to US$420 million. However, Nepal boasts a unique and diverse geological formation, from the subtropical jungle of the Terai to the high peaks of the Himalayas – eight of the world's ten highest mountains, including Mount Everest, are to be found there. Its culture is equally diverse; the dominant religions are Hinduism and Buddhism, and the population is made up of 12 major ethnic groups. It is not surprising, therefore, that since it first opened its borders to outsiders in the 1950s, not only has Nepal attracted visitors keen to experience its environment and culture but tourism has become an increasingly important element of the national economy. In 2004, for example, tourism generated US$230 million, representing 15 per cent of the country's foreign exchange earnings and 3.5 per cent of GDP.

Initially, few tourists were able to visit Nepal – in 1961, just 4,000 tourist arrivals were recorded. The first organized mountain trekking commenced in 1966 and, by the mid-1970s, arrivals had increased to over 100,000 a year. Since then, tourism has continued to grow, albeit erratically, given the ongoing political instability in the country. A peak was reached in 1999 when almost half a million tourists visited Nepal; in 2004, this figure had fallen back to 385,000 arrivals. It is also important to point out that around two-thirds of all tourists are regional visitors from India and Bangladesh.

One of the principal attractions of Nepal is, of course, the opportunity for mountain trekking – approximately 25 per cent of tourists participate in trekking, either independently or in organized groups. There are four main trekking regions, the most popular of which are the Annapurna Circuit area which attracts more than half of all trekkers, and the Everest base camp route which attracts around 20 per cent. The remaining treks take place on the Kanchenjunga route and in the more recently opened Mustang region. All of these areas benefit from some degree of protection or control. For example, the Mount Everest (Sagarmatha) National Park was established in 1976, while the designation of the Annapurna Conservation Area in 1984, and its subsequent success as a community-based conservation project, has received widespread acclaim. However, despite the protection afforded by these designations, all trekking areas suffer from a number of tourism-related impacts.

From a positive perspective, tourism has provided an important source of income and employment, particularly for communities alongside the popular routes where trekkers require accommodation and food. Nevertheless, relatively few communities benefit from tourism expenditure (the remote western regions of Nepal, for example, remain untouched by tourism), resulting in severe inequalities in the distribution of wealth. Moreover, employment opportunities in tourism have led to labour shortages in other areas and sectors, in particular in traditional agriculture. Greater attention, however, has been paid to the physical (negative) impacts of tourism in the Nepalese Himalaya, with all the major trekking areas suffering similar problems, including:

- *Deforestation*. The demand for firewood to support tourism-related cooking and heating needs has had a significant impact on forests and vegetation, while the building of lodges along trekking routes has led to a major increase in demand for timber (there are, for example, over 700 lodges and tea-rooms in the Annapurna area alone). Within the national parks and other controlled areas, tree felling is strictly controlled; this has, however, resulted in extensive deforestation beyond the boundaries of protected areas.

- *Litter*. The accumulation of litter alongside trails and around popular settlements and camp sites along the trails is a significant issue. Many studies draw attention to the huge volume of litter, much of it non-burnable or non-biodegradable, left by trekkers and mountaineering expeditions.
- *Trail damage*. The erosion of trails is a common problem throughout the trekking regions. This has been exacerbated by the rapid growth in the number of trekkers over the past two decades, particularly in the Annapurna region.
- *Pollution*. Toilets close to rivers and streams and the use of soaps and detergents for bathing or washing clothes and dishes in streams is a major cause of water pollution. This is exacerbated by the disposal of untreated human waste in rivers and streams.

Throughout the trekking areas in the Nepalese Himalaya, efforts are being made to address these impacts. Numerous schemes exist to control deforestation, to mange the collection of litter, to repair trails, and to introduce alternative energy sources for cooking and heating. The Annapurna Conservation Area, in particular, benefits from income derived from the fees that all trekkers are charged – this is channelled directly into conservation and other local projects. However, the need remains to balance the economic benefits of tourism with effective environmental management in order to prevent further damage to the fragile mountain ecology. Specifically, effective policies and controls are needed; for example, despite regulations requiring the use of kerosene for cooking, deforestation is still a problem in the Mustang region. Otherwise, the longer term environmental and social fabric of the region may be irreversibly compromised.

Sources: Nepal (2000); MacLellan *et al*. (2000).

Effect on population dynamics

The final environmental stress identified by the OECD is that associated with population dynamics. More specifically, the geographic flows of tourists, governed by a variety of factors that influence such flows (see Chapter 6) and manifested in significant seasonal increases in the population densities of destinations, may create a number of environmental impacts. The most evident of these is congestion and overcrowding experienced both at key sites or attractions and in resort areas in general, a subsequent effect of which may be greater physical damage, enhanced pollution levels and so on. Of equal, if not greater, importance, however, are the increased demands placed on natural resources. That is, the tourism industry competes with other sectors for scarce natural resources, such as land and water, and during the high season, tourism may place excessive strain on such resources. The demand for water is a particular problem. Tourists tend to consume a much higher quantity of water while on holiday than at home, both directly through, for example, regular showering or bathing, or indirectly through the expectation that sheets and towels will be laundered daily. Consequently, the tourism sector frequently consumes significantly more water than other local industries. In fact, one study has suggested that, in some less developed countries, 100 luxury hotel guests consume as much water in 55 days as 100 urban families consume in two years (Salem 1994). Similar demands may also be made on power supplies; electricity power cuts are not uncommon in some resort areas with limited generating capacity, necessitating hotels to operate their own back-up generators.

As noted above, the physical impacts of tourism are not always negative. In other words, it is important to recognize that the development of tourism may act as a catalyst for environmental protection and improvement. In particular, the designation of nature reserves, national parks, wildlife reserves and other categories of protected areas and landscapes is often, though not always, directly related to tourism development, while the expenditure on environmental improvements, such as the 'greening' of run-down areas, the cleaning and renovation of the built environment, or water quality improvement schemes may all be driven by the need to enhance the attraction of an area to tourists. Similarly, specific historic sites often benefit from tourism. For example, Angkor Wat in Cambodia which, since 1992, has been a UNESCO World Heritage Site, attracts about a million visitors a year and almost one-third of ticket revenues is spent on restoration work. The World Heritage Committee has also launched the World Heritage Sustainable Tourism Programme. Nevertheless, tourism remains a double-edged sword, with rapidly increasing numbers of visitors posing a potential threat to the physical fabric of the temple complex in the longer term (MacKenzie 2006).

Socio-cultural impacts

It has long been recognized that tourism has an impact on destination societies and cultures; indeed, some earlier tourism impact studies focused specifically on this topic (e.g. Smith 1977). It has also long been accepted that such impacts are likely to be more evident or keenly felt in tourist destinations in developing countries, where the difference in cultural and economic characteristics between local people and, primarily, relatively wealthy Western tourists is likely to be greatest (WTO 1981). At the same time, in some respects it would be considered unfortunate for tourism *not* to have some socio-cultural consequences on destinations; as a catalyst of development, tourism is usually promoted with the purpose of economic and social betterment. Moreover, tourism is seen by some as a means of achieving greater international harmony and understanding (WTO 1980) although, perhaps inevitably, it is the negative (and, frequently, emotive) socio-cultural impacts of tourism that attract most attention: 'tourists seem to be the incarnation of the materialism, philistinism and cultural homogenisation that is sweeping all before it in a converging world' (Macnaught 1982).

Although it is likely, if not inevitable, that destinations experience social and cultural change as a result of tourism development, the extent of that change is dependent on a number of factors. The socio-economic 'gulf' between tourists and local communities has already been mentioned but, in addition, the degree of socio-cultural change may be influenced by:

- *The types and numbers of tourists/tourist behaviour*. It is usually believed that higher numbers of mass tourists will impact more on host societies than smaller numbers of independent or responsible travellers, although smaller numbers of tourists in places that are relatively untouched by tourism may have a significant impact.

- *The size and structure of the tourism industry*. The larger the tourism industry relative to the local community, the greater its socio-cultural impact is likely to be.
- *The relative importance of the tourism industry*. The consequences of tourism will be more keenly felt in destinations that are highly dependent on tourism, although established resorts may have a variety of controls in place to limit such consequences.
- *The pace of tourism development*. Research has shown that socio-cultural impacts are more likely to be experienced when the development of tourism is rapid and uncontrolled.

It is also important to note, of course, that all societies and cultures are dynamic; they are all in a constant state of change and no society is immune from external influences (see Chapter 5). Tourism is undoubtedly one such influence but, frequently, tourism contributes towards, but does not cause, socio-cultural impacts and change. Nevertheless, tourism development is often blamed for what are seen as undesirable changes in destination societies, and therefore care must be taken to determine the precise components of socio-cultural change.

Tourist–host encounters

The basic context for the socio-cultural impacts of tourism is the so-called 'tourist–host encounter'. In other words, tourists inevitably come into contact with local people (hosts) in destination areas and, despite the usual brevity of such encounters, a variety of social processes are at work which determine the nature of that encounter (see Plate 7.3). This, in turn, goes some way towards determining or explaining the potential socio-cultural impacts of tourism because, generally speaking, the more unbalanced or unequal the encounter or relationship, the more likely it is that negative impacts will occur.

As Lea (1998) notes, there are potentially as many types of encounter as there are tourists and hosts. Nevertheless, a number of common characteristics can be identified with tourist–host encounters, foremost of which is the wealth of tourists relative to that of local people, particularly in developing countries. This may lead to feelings of resentment or inferiority on the part of local people. Four other principal characteristics are usually highlighted (Wall and Mathieson 2006: 223):

- Most encounters are transitory or fleeting, the resultant relationships being shallow and superficial.
- Most encounters are constrained temporally (the two weeks of the holiday or the tourist season) and spatially (the location or separateness of tourist facilities).
- Most encounters are pre-planned or lack spontaneity as hospitality becomes commercialized.
- Tourist–host encounters tend to be unbalanced, local people perhaps feeling inferior or subservient to tourists.

Plate 7.3 ***Argentina, Estancia Santa Susana, near Buenos Aires: Cultural performance at historic ranch.***

Source: Tom and Hazel Telfer.

Further characteristics may be added to this list, namely that many tourists travel with an apparent lack of knowledge, understanding or sensitivity to local culture or customs in destination areas. This represents another potential barrier to balanced or meaningful tourist–host encounters, and supports the argument for more responsible behaviour on the part of tourists (Chapter 6).

It is also important to point out that different categories of tourists come into contact with local communities in contexts other than the holiday. In other words, there are various ways in which mobile populations (including tourists) interact with 'static' communities (Hannam *et al.* 2006) and, as a result, increasing attention is being paid to the issue of resident tourists, second home owners and migrant workers who serve the needs of new tourist populations in destination areas. For example, it has been observed that conflicts exist between British migrants and local communities in southern Spain (O'Reilly 2003) while, more generally, complex relationships exist between local people and different types of tourist migrants (Williams and Hall 2000). However, here we are concerned with the more 'traditional' socio-cultural consequences of tourism.

The way in which local communities respond to the characteristics outlined above and, indeed, to the impacts of tourism in general is discussed shortly, but first, it is important to review the principal socio-impacts of tourism development. Typically, social and cultural impacts are referred to collectively, though it is useful to distinguish, albeit somewhat artificially, between social and cultural impacts. Thus, social impacts may be thought of as the more immediate effects of tourism on local people and their lifestyles, whereas cultural impacts refer to longer term changes that occur in the context of social values, attitudes and behaviour, as well as changes in the production and meaning of cultural art forms and practices.

Social impacts

From a positive perspective, tourism may have a variety of beneficial consequences for destination societies. These include infrastructural developments, the improvement of the physical environment and the provision of facilities that benefit tourists and local communities alike. More specifically, tourism provides employment opportunities which, in some countries, have brought a new freedom and independence, and improved social conditions, to many women. There are also other potential benefits. In The Gambia, for example, many schools benefit from charitable donations from tourists; in fact, on their return home, many visitors, establish small charities to raise funds for school buildings and materials, providing educational opportunities to many Gambian children (www.friendsofgovi.org.uk).

Conversely, tourism also has a number of less beneficial impacts on host societies. In general, for example, traditional community structures may be transformed as younger people are drawn from inland rural areas to the coast or cities to work in the tourism industry. Such migration patterns often result in a population imbalance in rural areas and the polarization of societies between younger, more affluent groups and older, more traditional generations. More specifically, the presence of tourists and their activities impact on local people in a number of ways:

- *The demonstration effect*. Tourism introduces new or alien values or lifestyles into destination areas. Local people may attempt to emulate behaviour or styles of dress, or strive to achieve levels of wealth demonstrated by tourists.
- *Crime*. Although there is little evidence to directly link increases in crime with the development of tourism, there is little doubt that where there are significant numbers of tourists criminal activity is also evident. This may result in increased expenditure on law enforcement, the growth in activities such as gambling and black market operations, increased crime against residents and, potentially, a reduction in tourism.
- *Religion*. In many tourism destinations, religious buildings, shrines and practices have become commoditized. That is, they have become an

attraction and a part of the tourism product, gazed upon and 'collected' by tourists. As a result, there is frequently conflict between local communities, devout visitors and tourists as religious rituals or places of worship are disrupted by tourists.

- *Prostitution/sex tourism.* Although care must be taken in apportioning the blame on tourism, there is no doubt that, in many destinations, such as Thailand, Cuba, Sri Lanka and the Philippines, tourism has led to an increase in prostitution. The social impact of this can be devastating, particularly in the case of child prostitution and the spread of sexually transmitted diseases (Ryan and Hall 2001; Bauer and McKercher 2003).

Cultural impacts

Over time, the culture of host societies may change and adapt either directly or indirectly as a result of tourism. Much of the literature is concerned, in particular, with the way in which cultural forms, such as arts and crafts or carnivals, festivals and religious events, become adapted, trivialized, packaged and commoditized for consumption by tourists, and there is certainly no doubt that this occurs on a wide scale. Many art forms become mass produced as souvenirs (i.e. 'airport art') while, frequently, cultural rituals are transformed and staged for tourists, becoming devoid of all meaning to the participants. For example, the style, production and cultural significance of the raksa dance masks in Sri Lanka have undergone significant change as a result of their production as souvenirs (see Box 7.2).

Box 7.2

The commoditization of dance masks, Sri Lanka

One of the most popular and widely available tourist souvenirs in Sri Lanka are colourful 'raksa' or 'devil dance' dance masks; at the same time they are also a striking example of the way in which cultural artefacts may be adapted and commoditized by tourism. Dance masks fulfil an important role in many forms of ritual and ceremony in Sri Lanka. Indeed in representing the images of a variety of deities and demons they have long been used in a number of different contexts, such as folk dramas, festivals and rituals of exorcism and healing. They have a fundamental meaning and significance in the performance of such rituals and, at the same time, the manufacture of the masks is also a recognized and socially important activity within the local community. However, the rapid growth of tourism to Sri Lanka during the late 1970s and early 1980s resulted in the masks being appropriated as an appealing and commercially attractive representation of Sinhalese culture for the tourist market. As they became mass produced, they were removed from the traditional system of production and performance and, as a consequence, not only did they lose their cultural authenticity but also the manufacturers of the masks suffered a decline in their social status. More specifically, dance masks which were manufactured for tourist consumption began to be produced in a variety of different sizes, decorations and colours to suit the tastes of tourists rather than in a more traditional or authentic style. At the same time, the manufacturers of masks (normally small family enterprises) sacrificed their traditional social role in the complex production–performance ritual as they began to mass produce inauthentic dance masks for commercial gain.

Source: Simpson (1993).

It is also true, however, that tourism also encourages the revitalization or resurgence of interest in traditional cultural practices and there are many examples of how tourism supports the redevelopment of traditional art forms and production techniques. For example, the traditional craft of greenstone carving in the southern Indian town of Mahabalipuram has been revitalized by the demand for souvenirs. The products, though intended for sale to tourists, are no less authentic than those produced through similar techniques hundreds of years ago

Less visibly, tourism also contributes to broader, deeper cultural transformations in destination societies. These are changes that occur in a society's values, moral codes, behavioural modes and identifying characteristics, such as dress and language. As mentioned earlier in this chapter as well as in Chapter 3 on globalization, it is difficult to separate the influence of tourism from other factors which induce cultural change but, nevertheless, it is generally accepted that tourism can accelerate this process, largely through what is known as acculturation.

Acculturation is the process whereby two cultures come into contact (for example, through tourist–host encounters) and over time they become more like each other through a process of borrowing. By implication, if one culture is stronger or more dominant than the other, it is more likely that this borrowing will be a one-way process. The extent to which tourism contributes to the acculturation process will vary depending on a variety of factors, such as the cultural gulf between the tourist and the host and the influence of other forces. In some cases cultures are more resilient (see Plate 7.4), yet, in many instances, cultural change can be directly attributable to tourism.

Plate 7.4 ***Indonesia, Bali: Traditional cultural ceremonies remain very important in Bali, even though it is a major tourism destination.***

Impacts on tourists

While most attention is focused on the impacts that tourists have on local communities, it is important to note that there exist a number of 'reverse' social impacts; that is, tourists themselves may experience impacts as a result of visiting tourist destinations. At a basic level, such impacts occur when tourists become the victims of crime or health problems; indeed, both have always been regarded as risks associated with travel and tourism. More recently, however, the opening up of more distant or exotic locations to mass tourism has resulted in increasing incidences of tourists contracting serious diseases such as malaria. Similarly, an increase in the number of cases of 'old' diseases, such as diphtheria, which have been virtually eradicated in Western societies, and of sexually transmitted diseases, has been directly associated with the growth in tourism to previously less accessible areas, particularly the transitional economies of Eastern Europe. However, a variety of other 'reverse' socio-cultural impacts may be identified, including:

- The internationalization of fashions and tastes in clothing, music, art forms, cuisine and so on.
- A reduction in national xenophobia resulting in a greater awareness and acceptance of different cultures and therefore tourism as a force for peace.
- The verification or changing of tourists' perceptions of different places, peoples and cultures; studies have also shown that tourists often return home with a more positive attitude towards their own society and culture.
- The adoption by tourists, either on a temporary or permanent basis, of new cultural practices.

The impacts of tourism: local community responses

Despite the usual focus on the impacts of tourism and tourists on destination environments and societies, local communities do not, of course, passively accept these impacts as an inevitable consequence of tourism development as seen in Chapter 5. Equally, local communities do not always respond to the impacts of tourism in ways which, from a Western-centric perspective, might be expected. For example, although the development of tourism in Cyprus has resulted in significant environmental degradation, this is not considered a problem by many Cypriots for whom entrepreneurialism and wealth creation for their families is a strong cultural value that outweighs environmental concerns (Sharpley 2001).

In other words, local communities respond to the development of tourism and its associated impacts in a variety of ways; these responses, in turn, may reflect local culture and values as well as more specific factors, including:

- The nature and scale of tourism development;
- The structure/ownership of the local tourism industry;
- The stage of development/maturity of the tourism sector;
- At an individual level, the degree of involvement in/benefit from tourism.

A number of models have been developed that explore local community responses and attitudes towards tourism development. These provide a useful framework for exploring ways of managing the impacts of tourism, and are considered in some detail in Wall and Mathieson (2006: 227–235). However, there are common elements to these models which may be summarized as follows. Typically, the models suggest a continuum of attitudes or behavioural responses to tourism development, from negative to positive. For example, in Doxey's (1975) widely cited framework, it is suggested that local residents in tourist destinations suffer varying degrees of irritation depending on the extent to which their lives are disrupted by tourism. These levels of irritation progress from euphoria, through apathy and irritation, to antagonism and, finally, to a resigned acceptance of tourism impacts. Doxey also suggests that the level of irritation will grow as the destination develops in a manner similar to Butler's (1980) resort life cycle; in other words, there is a unidirectional increase in irritation reflecting the stage of tourism development. This may not necessarily be the case, however.

For example, studies have shown that communities' responses to tourism's impacts may follow a continuum from welcome to rejection, though in different behavioral contexts (Figure 7.6). Research by Dogan (1989) and Ap and Crompton (1993) identifies a variety of responses, from favourable through to unfavourable, that are largely dependent upon the extent to which communities feel that the inconveniences or impacts of tourism are balanced by the benefits it brings.

Dogan (1989)	**Ap and Crompton (1993)**
Adoption: replacement of traditional host social structures with the adoption of tourist culture	*Embracement:* enthusiastic welcoming of tourists
Revitalization: tourism is used to preserve, promote or revitalize local culture for display to tourists	*Tolerance:* acceptance of the consequences of tourism in recognition of its benefits
Boundary maintenance: physical/social boundaries are established between tourists and local communities	*Adjustment:* alteration of behaviour to avoid the inconveniences resulting from tourism
Retreatism: local community avoids contact with tourists/develops an increased cultural consciousness	*Withdrawal:* physical or psychological distancing from tourism and tourists
Resistance: hostility and aggression against tourists/tourism industry	

Figure 7.6 *Community responses to tourism impacts.*

Similarly, other studies have looked at the differing attitudes towards tourism within communities, the purpose being to identify factors which may explain why different members of destination communities may respond in different ways at any point in time (or stage of tourism development). Typically and, perhaps, not surprisingly, these demonstrate that those members of local communities who are more involved in or economically dependent on tourism will view it more favourably or embrace tourism development more positively; conversely, those who are less dependent on tourism but, nevertheless, are affected by its impacts, will respond negatively. Research in Cyprus confirms this although, even among those involved in the tourism sector, resentment against tourists emerges when local tolerance levels are exceeded (Akis *et al.* 1996).

What these studies and, indeed, the analysis of tourism impacts in general demonstrate is the fact that the consequences of tourism development cannot be viewed from a simplistic, descriptive and implicitly Western-centric perspective. For tourism to contribute to sustainable development there is a need to manage local resources in a manner which reflects local values, the local political economy and local development needs. This is the focus of the final section of this chapter.

The impacts of tourism and sustainable development

Tourism must, of course, be planned and managed effectively in order to optimize its developmental potential. Tourism planning is concerned, as Wall and Mathieson note (2006: 293), with 'the process of making decisions about future desired states and how to attain them'; its focus tends to be longer term and, in the case of many developing countries, it is embodied in overall tourism master plans which provide a broad vision for the development of the destination as a whole. Many books address the tourism planning process, albeit from different perspectives (see e.g. Murphy 1985; Inskeep 1991; Hall 2000). Tourism management, in contrast, has a much shorter time scale and is concerned with processes and techniques designed to accommodate tourists and tourism development but, at the same time, to minimize their negative consequences. Such processes and techniques typically fall under two headings, namely: managing physical resources (Newsome *et al.* 2000), including land designation, spatial planning strategies and site management; and managing visitors. In both cases, a variety of management and regulatory measures are usually proposed.

An overview of tourism planning and management is beyond the scope of this chapter. The important point, however, is that if progress is to be made towards achieving sustainable development through tourism, effective means are required to assess or measure whether such progress is being made. In other words, while appropriate planning and management is, of course, a vital ingredient of the tourism development process, it is important to measure and monitor the extent to which tourism development is achieving local sustainability needs and objectives. Consequently, attention is now focused increasingly on a systems approach to tourism planning (see e.g. Tribe *et al.* 2000), a dynamic process that designs, implements, monitors and adapts tourism policy according to local developmental needs and resource limitations. Inherent in the systems approach is the requirement for establishing indicators of sustainability.

Sustainability indicators

Sustainability indicators provide a basis for monitoring and measuring the extent to which the key sustainable developmental issues of a tourism destination are being met. Examples include economic and social benefits, and particular resource use or tourism-specific goals, such as tourist satisfaction or seasonality. Not only do they help to clarify goals, they also provide a focus for identifying and assessing stresses on the environment, for measuring both the impacts of tourism and the effects of management actions. More broadly, sustainability indicators may fulfil the following purposes within the planning process (UNEP/WTO 2005: 73):

- they provide a baseline for measuring changes in the condition of resources and for assessing progress in satisfying local community needs;
- they represent a set of targets that form the basis of tourism development policies and actions;
- they provide a framework for assessing the effectiveness of actions;
- they enable the evaluation, review and modification of tourism development plans and policies.

Key to this process, of course, is the identification and selection of appropriate indicators, a task that ideally should be built into the process of local consultation and participation. The WTO (2004b) identifies numerous such indicators; however, these should be relevant to the local context, be easily measurable, and provide clear and credible information. In a recent publication (UNEP/WTO 2005), 12 baseline issues with corresponding baseline indicators are suggested. These are summarized in Figure 7.7.

In the context of measuring and managing tourism impacts, a fundamental requirement is the establishment of limits to environmental or socio-cultural change. The question is: How can those limits be established? Two concepts deserve consideration.

Carrying capacity

The concept of carrying capacity refers, quite simply, to the number of tourists that a destination or site can accommodate (or 'carry') without negative impacts on the local environment or society, or a reduction in the quality of the tourist experience. A number of different carrying capacities may be measured:

- *Physical capacity*. The actual number of tourists a place can physically accommodate.
- *Ecological capacity*. The extent to which the local ecology can withstand the impacts of tourism.
- *Socio-cultural capacity*. The limit of social or cultural impacts and change a local community will accept.
- *Psychological capacity*. The amount of congestion that will be tolerated by tourists before they feel their experience is being impaired.

Baseline issue	Baseline indicators
Local satisfaction with tourism	• Local population satisfaction levels
Effects of tourism on communities	• Ratio of tourists to local people at different periods • Recognition of tourism benefits for local communities (services/infrastructure)
Tourist satisfaction	• Levels of tourist satisfaction • Number/proportion of repeat visitors
Seasonality	• Arrivals by period • Occupancy levels by period • Proportion of tourism employment that is permanent/full-time
Economic benefits of tourism	• Numbers/proportion employed in tourism • Net economic benefits of tourism (income)
Energy consumption	• Per capita energy consumption • Proportion of energy from renewable sources
Water usage	• Water consumption per tourist/establishment • Water saving/recycling
Quality of drinking-water	• Proportion of establishments providing drinkable water • Number of water-related illnesses among tourists
Sewage treatment	• Number of tourist establishments treating sewage • Proportion of sewage per establishment being treated
Solid waste management	• Volume of waste produced • Volume of waste recycled
Development control	• Existence of land use/development policy • Proportion of land subject to development controls
Visitor management	• Total number of tourist arrivals • Density of tourist numbers at specific locations

Figure 7.7 ***Baseline issues and indicators for sustainable tourism development.***

Source: Adapted from UNEP/WTO (2005).

Carrying capacity remains the subject of intense debate in the tourism literature (McCool and Lime 2001). In particular, questions remain over how particular capacities are established, how they are measured and monitored, how they influence the tourist experience, what limits of damage or change are acceptable, and on what basis capacities are set. Nevertheless, carrying capacity does provide an appropriate measurement or indicator tool in particular circumstances. For example, in the late 1990s, the island of Malta revised its tourism development policy on a carrying capacity assessment linked to the supply of accommodation. It was decided that the development of tourism should be based on the then existing bed stock supply rather than on an expansion of the accommodation sector.

Limits of acceptable change

An alternative approach to establishing limits for environmental and socio-cultural change is the concept of Limits of Acceptable Change (LAC). Rather than focusing on the numbers of tourists or scale of activity, LAC recognizes that it is the impact or degree of change that is the problem. Therefore, LAC establishes limits to the impacts of tourism on destination environments and societies, limits that should be decided on by local consultation and measured by indicators. In this sense, LAC precedes or is inherent within the sustainability indicator process.

To summarize, then, the development of tourism is inevitably accompanied by impacts on destination environments and societies. In the context of international tourism, some of the impacts are generated by forces controlled by agents outside the destination country. The nature and extent of those impacts are determined by a variety of factors but, for tourism to contribute to the sustainable development of the destination area, negative impacts should be managed and controlled within local levels of environmental and social tolerance while positive impacts should be optimized. Contemporary approaches to this challenge favour the sustainability indicators approach although, as the final chapter of this book now considers, there is no single solution to the tourism development dilemma.

Discussion questions

1. **Is mass tourism the best development option to maximize the economic impact of tourism?**
2. **What are the challenges in using various measures to minimize negative impacts such as carrying capacity or sustainability indicators?**
3. **If a proposed tourism development in a remote location can generate a large number of new jobs, should it proceed if there will be negative social and environmental impacts?**
4. **If elements of a local culture can be preserved by putting them on display for tourists, should they be appropriated for the tourism market?**

Further reading

Holden, A. (2000) *Environment and Tourism*, London: Routledge. [Focusing specifically on the environmental consequences of tourism, this book provides an in-depth analysis of the nature and extent of, and potential solutions to, the impacts of tourism on the physical environment. Numerous relevant case studies are a particular feature of the book.]

Wall, G. and Mathieson, A. (2006) *Tourism: Change, Impacts and Opportunities*, Harlow: Pearson Education. [This book is a new, updated and revised edition of Mathieson and Wall's book, *Tourism: Economic, Physical and Social Impacts* (1982). Combining the original detailed analysis of tourism's impacts with additional sections addressing contemporary issues and challenges, this remains the most comprehensive book on the subject of the impacts of tourism.]

Websites

The World Tourism Organization website contains information relating to the Tourism Satellite Account: www.unwto.org/statistics/tsa/references/tsa_references.htm.

The United Nations Environment Programme website focuses on climate change. There are a variety of useful links on the page including one to a programme relating to development and climate. The programme acknowledges the pressing issues for many developing countries such as poverty, and food security, to name but a few: however, the project aims at identifying development paths linked to positive climate outcomes as well as facilitating dialogue and decision making with key stakeholders both nationally and internationally: www.unep.org/themes/climatechange/.

8 Conclusion: the tourism development dilemma

Learning objectives

When you have finished reading this chapter, you should be able to:

- **Be aware of the pros and cons of pursuing tourism as a development option;**
- **Be familiar with the debates surrounding the various paradigms of tourism development;**
- **Appreciate the range of controlling external forces such as the global market economy that may hinder or help development through tourism;**
- **Identify the potential conflicts and ways that multinational corporations, local élites and local residents can work together in tourism development.**

Tourism has increasingly become a favoured development tool in many developing countries. With the relative ease of entry into the tourism market and its purported ability to generate foreign exchange and create employment, it is no wonder that it is being pursued. However, like any development option or avenue of economic endeavour, it comes with a cost. This, then, is at the heart of the tourism development dilemma. Tourism represents an attractive, and perhaps the only, means of stimulating economic and social development for some developing nations. However, frequently that development either fails to materialize, benefits only the local élite or multinational corporations, or is achieved with a very high social, environmental or economic cost. In the developing world, tourism is usually implemented through a top-down planning approach, and decision making is 'predominately based on the interventions of government agencies and large tourism firms, resulting in the dominance of external, often foreign capital and the marginalisation of local people' (Liu and Wall 2006). Developing countries opting into the tourism industry will encounter both the positive and negative consequences of this globally competitive industry, and the challenge lies in accepting or managing the negative consequences in the hopes of obtaining the potential long-term benefits of tourism. The complexities of using tourism as a development tool and the dilemma that many countries face in coping with the uncertainty that tourism brings have been the focus of this book. The tourism development process intersects with the economic, political, environmental and social conditions in the destination and is also framed by the global political economy. The first section will focus on the development imperative and its relationship to tourism, and will examine the scope and realities of the developing world where tourism takes place. The second section will focus on the imperative of sustainability which has come to the forefront of the development debate. The tourism development dilemma will then be examined through a proposed framework. This framework covers current influences on tourism, the form of tourism that is developed, and the responses to tourism. The framework then focuses on the resulting trade-offs that are made in the tourism development process which are at times both competing and conflicting. These trade-offs illustrate the differences between tourism planning and management in developing countries with the idealism of sustainable development.

The development imperative and tourism

Box 1.1 contains the UN Millennium Development goals. These are global goals that, if reached by 2015, half a billion people will be lifted out of poverty and a further 250 million will no longer suffer from hunger. The goals are ambitious, a challenge to meet, and yet necessary targets. In describing the status of the world's population, Sachs (2005) places people on the 'ladder of economic development' with the higher rungs representing steps up the path to economic well-being. Approximately one billion people around the world or one-sixth of humanity are too ill, hungry or destitute to even place a foot on the first rung of the development ladder. They are the 'extreme poor' of the world and are fighting for survival. If they fall victim to a natural disaster (drought or flood), serious illness, or collapse of the world market price for their agricultural cash crop, the result will be extreme suffering and perhaps death (Sachs 2005). A few steps up the ladder are approximately 1.5 billion people who are at the upper end of the low-income world. Sachs (2005) describes these people as 'the poor', living above mere subsistence and, while daily survival is virtually assured, they struggle to make ends meet in cities and rural areas. They face chronic financial hardship and lack basic amenities including safe drinking water and functioning latrines. Together 'the poor' and 'the extreme poor' make up approximately 40 per cent of humanity. Further up the ladder are another 2.5 billion people in the middle-income world earning a few thousand dollars per year; however, they are not to be confused with the middle class in rich countries. Most live in cities, have adequate clothing, and their children go to school. They may have some amenities in their homes such as indoor plumbing, and they may be able to purchase some means of transportation, such as a scooter and later possibly an automobile (Sachs 2005). They have adequate food although some are following the trend in developed countries of eating fast food. The remaining one billion people, or one-sixth of humanity, are higher up the development ladder and are in the high-income world. The people in the high-income households include the approximately one billion in the rich countries but also a growing number of affluent people living in middle-income countries. There are tens of millions of high-income people in cities such as Shanghai, São Paolo or Mexico City (Sachs 2005). On the positive side, Sachs (2005) states that more than half the world is climbing the development ladder, as is evident through measures of economic well-being such as increases in income, life expectancy, education, access to water and sanitation, and falling infant mortality rates. However, one-sixth of humanity is not even on the development ladder and caught in a poverty trap. From a basic needs perspective, the Food and Agriculture Organisation of the United Nations (FAO) released its annual report entitled *The State of Food Insecurity in the World* in October 2006; the report indicated that there are 820 million people now going hungry in developing countries, which is more than there were in 1996. In exploring pro-poor tourism, Chok *et al.* (2007) refer to the UN-Habitat (2003) report on slums which indicates that the urban population in developing countries is expected to double to four billion in 30 years' time and currently almost 80 per cent of the urban population in the world's 30 least developed nations live in slums. It is the people of the developing world with whom the well-off, domestic and foreign tourists interact. It is then easy to see how the simple haggling over the price of a souvenir in a local market in a developing country can have profound repercussions for someone struggling to make ends meet.

It is important to also recognize that those who are disadvantaged in the labour market must not be thought of as passive acceptors of their fate (Potter *et al.* 1999). Low-income households display a wide range of coping mechanisms (Rakodi 1995 cited by Potter *et al.* 1995). Rakodi (1995) identifies three main categories for urban households' strategics for coping with worsening poverty. Under changing household composition the strategies include migration, increasing household size to maximize earning opportunities or not increasing household size through fertility controls. Under consumption controls Rakodi (1995) lists reducing consumption, purchasing cheaper items, removing children from school, putting off medial treatment, delaying repairs to property, and limiting social contacts including rural visits. The final category of strategies relates to increasing assets. These include having more household members in the workforce, starting up enterprises where possible, increasing subsistence activity such as growing food or gathering fuel, and increasing scavenging and sub-letting rooms and/or shacks. Rakodi (1995) states that while not all of these strategies are available to all households, they should be linked to policy responses. In the context of these strategies, how would a member of a disadvantaged household respond to tourism development in an urban area? Selling souvenirs, acting as an informal tour guide, or moving to a new tourist area are just a few of the options relating to tourism. In a study of tourism employment in Bali, Indonesia, Cukier (2002) found that most informal sector workers were not marginalized but rather were earning above minimum wages and often more than formal front desk employees. These strategies listed by Rakodi (1995), along with Cukier's findings, reflect the importance of small enterprises, the informal sector and local entrepreneurial activity.

The contribution of tourism also needs to be considered within the wider development issues that any specific country may be facing. Writing on the topic of nature tourism and climate change in southern Africa, Preston-Whyte and Watson (2005: 141) made the following comment revealing the complexity of development:

> At present the major challenge facing southern African governments is to deal with the colonial land legacy, HIV/AIDS and associated poverty so that the region's inhabitants will find themselves in a position to actually benefit from the potentially favourable effects of climate change on nature tourism.

The evolution of development theory and tourism

The challenges of the developing world are complex, and the search continues for strategies, programmes and initiatives from within and outside the developing world to help improve the situation in developing countries and overcome barriers to move up the development ladder. As illustrated in Chapter 1, the concept of development has changed over time and has been a source of controversy. There has been a shift from top-down economic models to more broad-based approaches that are bottom-up and focus on satisfying basic needs and the notion of sustainability. How development is measured has also moved beyond only economic indicators such as GNP/capita to more broad-based indicators such as the UNDP Human Development Index. Box 8.1 includes a discussion of the relationship

Box 8.1

Tourism development and human rights

The concept of human rights intersects with tourism and development issues on a number of different fronts and it is useful to examine tourism within the context of the *Universal Declaration of Human Rights* (Hashimoto 2004). Tourism Concern, for example, is promoting a boycott on travel to Burma, in part due to human rights violations. The United Nations (2007) adopted the Universal Declaration of Human Rights in December 1948. Other, more specialized human rights conventions have also been developed, such as *The International Covenant on Economic, Social and Cultural Rights* in 1966 and *The Convention on the Rights of the Child* in 1989 (Freeman 2005). With the notion of development broadening out from an economic focus to a more 'human development' approach, human rights and development may be seen as conceptually overlapping (Freeman 2005). As an example of this, the United Nations Development Programme has a section on the protection of human rights in its policies (Freeman 2005). Sen (1999) argues from the perspective of 'development as freedom' and notes five specific types of rights and opportunities that can help to advance the general capability of a person. The five rights and opportunities include political freedoms, economic facilities, social opportunities, transparency guarantees and protective security. The focus here is on selected issues that arise when considering tourism in the context of the Universal Declaration of Human Rights as suggested by Atsuko Hashimoto (2004). The Declaration has 30 articles and all can be examined in the context of tourism. A full discussion of all 30 articles is beyond the scope of this book; however, a few are included below for discussion purposes. The selection of the articles is not meant to imply that they are of greater importance than any of the other articles in the Declaration.

Article 1
All human beings are born free and equal in dignity and rights. They are endowed with reason and conscience and should act towards one another in a spirit of brotherhood.

- Tourism presents the opportunities for the understanding of different cultures.

Article 4
No one shall be held in slavery or servitude; slavery and the slave trade shall be prohibited in all their forms.

- The sex tourism industry can result in a form of slavery and slave trade.
- Forced labour including child labour in tourism.
- Tourists travelling to purchase the freedom of slaves.

Article 13, Parts 1 and 2
Everyone has the right to freedom of movement and residence within the borders of each state; and everyone has the right to leave any country, including his own, and to return to his country.

- In some countries the movement of nationals and tourists is restricted.
- In some areas locals are not permitted to go to certain locations, such as beachfronts, where resorts have been built.
- Some communities have been relocated to make way for tourism developments and not allowed to return.

Article 17, Parts 1 and 2
Everyone has the right to own property alone as well as in association with others; no one shall be arbitrarily deprived of his property.

- Tourism development may displace people to make way for hotels, golf-courses and tourism-related infrastructure.

Article 22
Everyone, as a member of society, has the right to social security and is entitled to realization, through national effort and international cooperation and in accordance with the organization and resources of each state, of the economic, social and cultural rights indispensable for his dignity and the free development of his personality.

- Tourism can commodify indigenous cultures to be sold as a tourism product.
- The promotion of cultural tourism may enhance and strengthen local cultures.

Article 23, Part 1
Everyone has the right to work, to free choice of employment, to just and favourable conditions of work and to protection against unemployment.

- Tourism generates employment but the quality of the work needs to be considered as well.

The few selected examples given above illustrate how tourism can be discussed within the context of the Universal Declaration of Human Rights. The edited work by Andreassen and Marks (2006) further explores the concepts of the 'human rights-based approach to development' and the 'human right to development'.

Sources: Hashimoto (2004, personal communication, Human Rights and Tourism Lecture, Brock University); Sen (1999); UN (2007).

between tourism and human rights, which has been brought into the broader scope of development. Telfer (2002a, in press) summarized these changes in development thinking over time and their influence on tourism development in terms of modernization, dependency, economic neoliberalism, alternative development and 'beyond the impasse, the search for a new paradigm?' (see Chapter 1). Initially in the 1960s, the focus of tourism under modernization was primarily economic with the belief that tourism generated increases in foreign exchange, employment and engendered a large multiplier effect, stimulating the local economy. In time the benefits of tourism were questioned, due to high rates of leakages and lower than expected multipliers. The negative impacts of tourism in developing countries were documented, paralleling the dependency critique of modernization. In the 1980s and 1990s the economic neoliberal paradigm gained prominence with a focus on international markets and globalization. Multinational tourism corporations extended their operations around the world, looking for locations with attractive destinations and lower production costs. Within the alternative development paradigm, sustainable tourism development has come to the forefront (Telfer 2002). The 1980s also saw the beginnings of an impasse in development thinking as the previous development paradigms could not explain all of the difficulties that developing countries were encountering (Schuurman 1996). As an example of a critic of globalization, Saul (2005: 3) argues:

> that very clear idea of globalisation is now slipping away. Much of it is already gone. Parts of it will probably remain. The field is crowded with other competing ideas, ideologies and influences ranging from positive to catastrophic. In this atmosphere of confusion, we can't be sure what is coming next, although we could most certainly influence the outcome.

A diversity of approaches has been put forward due in part to the criticisms over development paradigms of the past and few of these ideals are identified in Table 1.4. The diversity of approaches includes, but is not limited to, calls for an end to 'development', a new role for the state and the importance of civil society, social capital, transnational social movements, cultural studies, and development and security. These approaches have been raised in tourism through a diversity of areas including the importance of state involvement in tourism policy and for local participation in tourism, the desire to understand different worldviews in tourism, and the need for security for tourism to operate successfully.

The continued presence of poverty and inequality raises concerns over the development process and globalization (Sachs 1996; Saul 2005). Disillusionment with attempts at grand theory building is coming from a variety of perspectives including post-structuralism which originated in linguistics and philosophy: 'Post-structuralists question the epistemological basis and claims of the great theoretical approaches or "meta-narratives" such as liberalism, Marxism, or indeed "modernisation" ' (Randall 2005). McMichael (2004) highlights the controversy surrounding development and its shortcomings. With the failure of many countries in fulfilling the promise of development and a growing awareness of environmental limits there has been a re-evaluation of development. McMichael (2004) indicates that there have been two main responses. One advocates a thoroughly global market expanding trade and spreading wealth while the other re-evaluates the economic emphasis and moves to recover a sense of cultural community. Both of these trends are evident in tourism through the globalization of the tourism industry, by multinational hotel chains at one end of the spectrum and locally controlled community-based tourism at the other. Some development critics, however, have taken a more radical perspective, calling for the demise of development, including Sachs (1996: 1), who stated that 'development stands like a ruin in the intellectual landscape. Delusion and disappointment, failures and crimes have been the steady companions of development and they all tell a common story: it did not work.' In response to this, Thomas (2000) claims that although important points have been raised by critics about development from a radical position, they are not arguments about abandoning the concept. Deciding not to use the term will not solve the problems of such issues as poverty and powerlessness, environmental degradation and social disorder (Thomas 2000). Thomas (2000) suggests that post-development writers do not deny the need for change. What they are arguing for is that if change is to be done differently, it needs to be conceived literally in the context of different terms. Thomas (2000) concludes his remarks by quoting Robert Chambers (1997: 9) in response to Sachs' comments listed above: 'That is no grounds for pessimism. Much can grow on and out of ruin. Past errors as well as achievements contribute to current learning.'

The question for this book is: What can tourism do to help people, groups, communities, regions or countries climb the ladder of economic development (see Plate 8.1)? Development is defined in Chapter 1 as 'the continuous and positive change in the economic, social, political and cultural dimensions of the human condition guided by the principle of freedom of choice and limited by the capacity of the environment to sustain such change'. Tourism is only one industry and, while it is the focus of this book, it cannot possibly be the sole solution to the challenges of developing countries. If adopted, tourism needs to be viewed as part of a broader based approach to development in conjunction with other economic activities. Along with the UN Millennium goals there are also national development goals, and if we continue down in scale we travel through regions, cities, towns, villages, communities, and eventually we get to the individual. What opportunities does an individual have to take part if he or she so desires, and to participate directly or indirectly in the economy of tourism? Can they take up a position in the formal sector such as a front desk position at a beach resort in Sri Lanka or can they enter the informal sector and perhaps sell souvenirs in the streets of Cairo? At the community level individuals may come together perhaps with the help of an NGO or local government to launch a village tourism initiative such as the Kasongan pottery village in Central Java, Indonesia. At a regional or national level, the goal may be to establish tourism to attract visitor numbers, development charges and taxes, which can be used for broader development goals so that the nation can climb further up the development ladder.

Tourism is a complex, evolving, dynamic and sometimes volatile industry, and judging its role in the development process is not an easy task, as highlighted in this further example. An initial visit to the north-west coast of the island of Lombok,

Plate 8.1 ***Indonesia, Lombok: Local village very close to main tourist resort area of Sengiggi Beach.***

Indonesia, in 1994 saw an open beach with no tourism development. A year later a return visit to the area saw the same beach front property bounded by a large fence with a sign advertising a new hotel under construction. Access was now restricted. Time has passed, and now in place of the fence stands a new hotel; others have followed. How has the community changed? Do the locals benefit from the hotel? Is this particular hotel contributing to the development of the island? Is the hotel in a small way helping Indonesia to climb the development ladder? Is this hotel one piece of a much larger puzzle helping to contribute to the United Nations Millennium Development goals? All of these questions are difficult to answer, yet considering them opens the door to exploring the complex relationship that exists between tourism and development.

Development goals need to be framed within some hard questions – development for whom and by whom? We need to consider the priorities of governments, NGOs, private businesses (both large and small), communities and individuals. How do their goals relate to broader development goals? How do their priorities fit in with the global market and external political forces as well conditions in the destination? As noted in Chapter 5, *fakalakalaka* is the Tonga notion of how to be modern and how to do development (Horan 2002), and does not necessarily coincide with the macroeconomic indices used by Western international lending agencies. Based on the work by Iliau (1997), Horan (2002: 216) states that *fakalakalaka*

> was a more encompassing and in-depth concept than merely the personal gain of material possession, let alone just making money by selling tourists textiles. This notion of development is about total development of the individual: physical, spiritual and mental. Development then extends to all areas of life including personal relationships, family dynamics and the development of the community.

Beyond issues of poverty, indigenous communities such as those found in Tonga may, in fact, have differing perspectives of a development ladder identified by Sachs (2005) above. Clearly tourism which follows different development paradigms or schools of thought will have a different emphasis and, as noted in Chapters 3 and 4, it is important to understand the values, ideologies and strategies of these paradigms as well as who has the power and control to implement and/or enforce them.

The sustainability imperative and tourism

The above section has stressed the development imperative and how tourism is used as a development tool in the context of development paradigms. Within the broader development imperative and the overall need for improvements in developing countries, there is also the sustainability imperative. With the recognition that resources are limited and need to be protected for future use, sustainability has become a guiding framework for development. The sole focus on economic growth has been challenged by the more holistic approach theorized under sustainable development encompassing environmental, social and economic concerns. There is increasing recognition that local communities need to be brought into the planning process of tourism. However, Rogerson (2006) argues that in the past, sustainability focused on environmental sustainability and it is only recently that the focus has shifted towards poverty alleviation, as outlined in the UN Millennium Development goals. The evolution of sustainability and the challenges associated with the term have been explored in Chapter 2. Liu and Wall (2006: 160) state that:

> while tourism has gained prominence in the social and economic agendas of all levels of government and academics continually espouse the need to involve local people, sufficient attention is rarely accorded to the means by which the capabilities of local people to respond to tourism opportunities can be enhanced.

Redclift (2000) outlined five spheres of environmental activity in which to analyse sustainability. They are presented in Table 8.1 along with a series of questions to consider in terms of tourism development. The spheres are all highly interrelated but the questions posed concern the interrelationships between tourism and sustainable development.

Proops and Wilkinson (2000) explore the relationship between sustainability, knowledge, ethics and the law. They outline four areas of knowledge and understanding that are necessary to establish well-formulated policies for sustainability. These four areas for sustainability are adapted here to tourism, with cautionary notes with regard to the tourism development dilemma. The first is an

Table 8.1 Spheres of environmental activity and questions for sustainable tourism development

Sphere of production	• What are the environmental impacts of the production of tourism? • What are the effects of employment in tourism on health, welfare and community life? • How does tourism contribute to income levels? • What are the associated risks with the tourism industry? • What are the indirect consequences of tourism production activities such as waste, toxins and pollution?
Sphere of consumption	• How do different tourists consume tourism differently? • What risks (i.e. health, food) are there in the consumption of tourism? • What are the indirect consequences of tourism consumption such as food miles, ghost acres, ecological footprints, demonstration effect and commodification of culture? • What energy is generated to meet tourists' demands and the disposal of waste packaging to meet these demands?
Sphere of social capital/infrastructure	• What is the built environment for tourism? • What utilities are required for tourism (energy, water and waste disposal)? • How are the activities linked by transportation? • What public services (public parks, open areas, recreational services) are used by tourists?
Sphere of nature	• How are the countryside, forests and landscape used by tourists? • Do tourists and locals have access to natural areas (beach, wilderness)? • Are animal rights and welfare protected in tourism?
Sphere of physical sustainability	• What is the environmental quality for both tourists and residents? • What are the background processes that govern environmental quality for tourists and residents such as climate change, air pollution, ozone depletion, destruction of forests, and watershed basins due to anthropogenic causes?

Source: after Redclift (2000).

understanding of the natural world and how the activities of production and consumption impact on it. In the context of the developing world it is important for destinations to understand the power structures of the international tourism industry. The invitation to foreign-controlled companies to operate in the destination can leave the host country in a very dependent position. The creation of beach resorts or community-based village tourism will have different forms and attract different types of consumers. The understanding of the interaction of the built form with consumers in the destination and the resulting impact on the environment may be difficult to manage over longer periods of time.

The second issue is an understanding of human perceptions and motivations so that we can know why people indulge in behaviour that is destructive to nature. The tourists themselves, as consumers, bring with them not only their home culture but they enter the realm of tourist culture. Different tourists will have different values, behaviours and disposable income. Some tourists may be willing to adopt more environmentally friendly forms of behaviour; however, for many, travelling is a holiday and they may be more interested in relaxing than following restrictive codes of conduct.

The third area of knowledge and understanding necessary to establish well-formulated policies for sustainability is an understanding of ethical systems so that we can establish if human motivations that are destructive to nature might be morally constrained. There have been attempts by different organizations to develop codes of conduct for tourists such as the one presented in Figure 6.3. However, as Pearce (2005) argues, it is not clear from an evidence-based perspective that all suggested behaviours in codes of conduct will result in sustainable outcomes. An example of this (Pearce 2005) is the appropriateness of encouraging tourists to boycott locations with poor human rights records. He indicates that an alternative suggestion for tourists and the governments who represent them may be forces for social change in these communities. The second problem Pearce (2005) identifies with codes of conduct is that in some prescribed roles it is difficult to make the necessary judgements. One example given in this situation is the proposed boycott of a business. It may appear that a company is exploiting its workers with low pay and so the code of conduct will call for a boycott. However, the amount of money being paid may seem trivial in a developed country context, yet it may be 'substantial in Indonesia, and jobs that look menial and almost pointless in populated destinations such as India or China may provide dignity and some income of value to poverty stricken regions' (Pearce 2005: 23).

Finally, Proops and Wilkinson (2000) state that we need to understand the effectiveness of various systems of incentives and restrictions on human actions so that appropriate measures can be made law. Governments may enact policies and laws to regulate the tourism industry. Alternatively, governments may pro-actively offer investment incentives to attract developers. In a globalized economy, a developer may choose to go where the incentives are higher and the environmental and social regulations are lower.

Having presented these four statements, Proops and Wilkinson (2000) then argue that the statements are very problematic and impose significant difficulties in trying to establish a sustainable world. The difficulty with all of these statements is related to the complexity of the interactions of humans with the environment as well as the difficulty in understanding human behaviour towards the environment. The tourism examples provided above, in the context of the four statements, clearly illustrate the challenges associated with sustainable tourism development. It is important to also keep in mind that sustainability itself has come under a variety of criticisms including its being a Western notion of development. While it may not be possible for tourism to be completely sustainable, making all forms of tourism more sustainable becomes the challenge. The chapter now turns to explore a framework for understanding the tourism development dilemma.

The tourism development dilemma framework

In order to better assess the tourism development dilemma, a framework is presented in Figure 8.1 (see p. 226). The framework is based on the work of Potter (1995) in studying urbanization in small island Caribbean states. The framework is briefly outlined before the various components are described in more detail. In the centre of the framework is the tourism destination. It is presented based on the shape of an island, as the original framework was based on a hypothetical Caribbean island; however, the main concepts are transferable to other locations and scales. Some of the main influences on the tourism development process include modernization, globalization, production, dependency, postmodernism, consumption and sustainable development, and these issues have been explored throughout this book. These influences can have a major impact on how tourism develops in a destination. They may come from outside the destination, from within the destination, or both. It is important to recognize that this list of influences is not comprehensive and may differ from destination to destination. While the influences are discussed individually below, they are not isolated and are very much interrelated as they interact and overlap in the destination. Within the context of these influences, local and external tourism development agents take advantage of opportunities to develop tourism resulting in built form and reactions by individuals, groups, communities, regions and governments. The built form and the type of tourism developed, such as mass beach tourism or remote village tourism, will have an impact on the contribution of tourism to development.

Tourism not only has the potential to contribute to broader based development goals but, as illustrated in Figure 8.1, the resulting interactions have a geographic dynamic. In the context of the Caribbean islands such as Barbados, St Lucia and Grenada, tourism has obviously focused development and change in coastal areas and further development of the industry has occurred along the pre-existing coastal urban zone (Potter 1995). While the main tourism zone develops and expands, satellite resorts or attractions such as village tourism or ecotourism will open up if there are adequate transport links as indicated in Figure 8.1. In the case of ecotourism resorts, the periphery may become more significant than the initial main tourism resort area. The geographic location of the built form will also influence the extent to which tourism can contribute towards development. Tourism concentration will result in migration. The final section of Figure 8.1 is the trade-offs which are at the heart of the tourism development dilemma. The outcomes are listed under the four headings of economic, environment, social/cultural and political; however, the trade-offs may occur within and between these categories. Of course, under sustainable development the goal is to minimize negative impacts. In an ideal world, however, trade-offs may still occur. It is important to consider who makes the decisions as to what the trade-offs will be. Figure 8.1 not only explores the tourism development dilemma, it also draws together the material presented in previous chapters.

When examining Figure 8.1 it is important to keep in mind the issues of power and control. In Chapter 4 the tourism development process was examined and the first items to be explored were the values, ideologies, goals, priorities, strategies and resources of tourism development agents. There are a range of competing forces in tourism and the values of those in power will have a strong influence on the overall development of tourism and therefore will influence the potential to contribute to broader notions of development of the destination. Multinationals, international funding agencies, governments, NGOs, communities and the domestic private sector are a few of the many organizations influencing tourism development. As Bianchi (2002: 268) argues:

> the attempt to conceive of market behaviour in isolation from the ideologies and values of the different actors and interest groups involved, as reflected in the free market notion of *comparative advantage*, underplays the political nature of markets whereby the state has historically conditioned the activates of economic classes and, furthermore, ignores the uneven consequences of unlimited market competition.

Bianchi (2002: 268) focuses on the political economy of tourism which is summarized as 'the examination of the systemic sources of power which both reflect and constitute the competition for resources and the manipulation of scarcity, in the context of converting people, places and histories into objects of tourism'. When examining Figure 8.1 it is essential to consider the influence of power and control by various groups and individuals in the tourism development process.

Influences

Some of the main influences on tourism development are listed in Figure 8.1. The aim here is to examine these influences in terms of the tourism development dilemma and how they may have an impact on tourism being a force for overall development. Developing countries face some incredible challenges and strong forces from both beyond and within their borders. Although the influences listed below are considered individually for the purposes of analysis, many are highly connected and influence each other either directly or indirectly. They are presented here for discussion purposes to explore the range of influences interacting with tourism and destinations in a very dynamic global environment.

Modernization

Modernization is socio-economic development which follows an evolutionary path from a traditional society to one that is modern such as in Western Europe or North America (Schmidt 1989). Rostow (1967), for example, argued that for development to occur it had to pass through a series of stages from traditional society to the age of high mass consumption. Some would argue that the goal of development then is to become a modern state. There is pressure for developing countries to climb the ladder of economic development (Sachs 2005). Tourism then becomes one tool to help generate the resources to move in that direction. If maximum financial gain is the target to achieve this goal, large-scale mass tourism may be used. As indicated in Chapter 1, there are links between modernization and regional economic development theory, and one such theory emphasizes growth-poles. It could be argued that large-scale resort complexes such as Cancún in

Mexico and Nusa Dua in Bali, Indonesia act as growth-poles for the economy. Tourism development corporations may be established by governments to identify areas for large-scale development and then provide services to attract developers. There is also an interesting dilemma within modernization as it relates to tourism. Tourism brings with it conflicting pressures. One end of the spectrum stresses the importance of moving ahead to provide all the luxuries tourists demand and at the other end there are demands that some communities need to stay as they are so that tourists can see the ways of the past or experience the undeveloped world. Can tourism dictate to some communities that they need to remain in a non-modern state to attract tourists?

Globalization

As outlined in Chapter 3, globalization has facilitated the movement of goods, people, information, values and finances across political boundaries. In this era of free trade, global markets and economic neoliberalism, there is a great deal of pressure for destinations to open up their borders to trade, including tourism. At times this has resulted in the reduction of the power of the state and so, to a large extent, the multinational corporations hold a great deal of power in the tourism decision-making process. Globalization has changed the nature of sovereignty by creating overlapping levels of governance. For example, states must not only look after their own internal affairs as they open up their borders; they must also deal with international trade organizations, international funding agencies or, perhaps, regional trading blocs. Multinational tourism corporations are looking for good locations with favourable trade policies and low production costs to open their businesses. Developing countries often then compete to attract developers for tourism. Rodrik (2005) identifies the 'Political Trilemma of the Global Economy'. He suggests that (1) the nation-state system, (2) deep economic integration, and (3) democracy (mass politics) are mutually incompatible. At most, one could obtain two out of the three; if a country wanted to push for global economic integration, a choice would have to be made to give up on the power of the nation state or democracy (mass politics). Democratic politics would be reduced, since once countries join the global economy they must follow the financial rules set by the global market, leaving very little room for democratic debate over national economic policy making. In the case of tourism, joining the global economy may result in either a reduction in the power of the state as power shifts to multinationals or a reduction in democracy (mass politics). Once borders are open and restrictions lifted, there are criticisms that multinational corporations such as hotels, airlines or tour operators will function with high levels of leakage.

Another aspect of globalization is the international sources of funding. If developing countries want to attract tourism and they do not have the necessary funds (say for infrastructure), where can they turn? Easterly (2005) examined the complex process that a developing country has to go through to obtain aid money, such as preparing a 'Poverty Reduction Strategy Paper' that would go forward to the World Bank for evaluation. As is evident in the following statement, Easterly (2005: 187–188) is highly critical of the bureaucracy relating to obtaining and receiving foreign aid:

> Frontline staff in aid agencies can barely keep up with the flood of mandates, political pressures, procedural requirements and authorizations required to keep the aid money flowing. In the recipient governments, the relative burden on the management is even greater given the scarcity of skills, the extreme political pressures in factionalized societies and the prevalence of red tape and corruption.

He goes on to argue that foreign aid is continuing to lose support in rich countries just as liberal political and economic ideals are losing support in poor countries. If a developing country includes tourism as part of its strategy for development in an application for foreign aid, then the above comment by Easterly (2005) highlights the challenges ahead. International loans have also been criticized in the past for the accompanying required changes to domestic policy, which have had harsh impacts on recipient countries.

Globalization has a number of positive aspects for developing countries. Multinationals, for example, bring capital, knowledge transfer, business-operating systems and technology as well as stimulating entrepreneurial development. A second generation of multinationals is emerging, based in developing countries and expanding beyond their borders. A rapidly emerging aspect of globalization is technological change and information technology. The internet has allowed small operators in developing countries to advertise their products online. It has opened the door to new markets and offered a further avenue of power to tourism in developing countries for those able to access it. While technology has facilitated online bookings for companies, it has also empowered individuals through such aids as mobile phones and handheld portable translators. The experience of sitting in a restaurant in Indonesia with an informal sector tour guide and watching him use a handheld translator to communicate in English, Indonesian and Japanese with the people he is leading illustrates how technology can open up new opportunities if one has the funds to access the technology. This also illustrates the globalization process in culture.

Production

The factors of production in tourism will depend on the type of tourism developed and who is in control of the tourism development process. In community-based tourism such as in the agritourism complex in the village of Bangunkerto (central Java), Indonesia, the main inputs into the project were community participation and some governmental assistance. In order to pave the roads in the local village so that tour buses could get through, the entire village was out working. The men collected rocks from the site of a past volcanic eruption to use as part of the new road base while the women arranged the actual rocks and created the road bed in the village (Telfer 2000). In the context of large-scale tourism developments, criticism focuses on the issue that the factors of production may be imported and profits repatriated. However, there is evidence that large hotels can adopt policies of using local products to increase the local multiplier effect (Telfer and Wall 2000).

Potter (1995: 334) argues that 'globalisation is not bringing about a uniform world, rather leading to new and highly differentiated localities and regions'. He refers to this as part of the concept of divergence. In looking at the history of urbanization in the Caribbean, Potter (1995) states that development tends to be focused on the pre-existing core urban areas where infrastructure is already in place. Second, development has continued in a top-down, centre-out manner. Making comparisons to the tourism industry in the Caribbean, Potter (1995) indicates that tourism has continued to focus on coastal areas along with further development of pre-existing coastal urban zones. Over time, new extensions to coastal development have occurred in places such as Montego Bay and Ocho Rios. This concentration has geographic implications, as reflected in Figure 8.1. The larger resorts continue to evolve along the coast and newer resort areas open up over time. The clustering of resorts will draw not only tourists but also migrants looking for work, resulting in regional imbalances in the country. Potter (1995) also comments that the flow of workers in the tourism sector dampens the availability of labour for agriculture and may lead directly to the creation of idle land.

A final comment that relates to production is in the area of sustainability. There are increasing pressures for companies to operate in a more sustainable manner. More and more companies are publishing statements of corporate social responsibility or developing environmental programmes. Some tour operators have identified this as part of their marketing strategy and offer more responsible travel. Programmes such as Green Globe and the European Blue Flag campaign for beaches all reflect this trend. If these policies and programmes are followed, they represent opportunities to enhance the contribution that tourism can make to development as well as protecting the environment.

Dependency

Problems of dependency relate to external control of the industry as well as an over-reliance on tourism. Developing countries may have little alternative but to rely heavily on the tourism sector and national priorities may take a back seat to the priorities of multinationals. Tour operators, for example, hold a great deal of power and can quite quickly decide to change destinations for the year, leaving hotel operators trying to fill their rooms. In some developing countries national-based hotel chains have opened up, thereby reducing the level of external dependency. The problem of being over-reliant on one industry applies not only to tourism but to any industry. If there is a downturn in the market or increased competition, the industry will suffer. For example, if a tourist from Canada wants to go to the Caribbean to escape the cold winter, they have a wide range of destinations to choose from in the Caribbean. Political stability, safety and health concerns are all vital to the image of the destination; any negative event will have a large impact on tourism in the destination area. Tourism Concern, a UK-based NGO, is currently running a campaign entitled 'Sun, Sand, Sea and Sweatshops' which reveals that many hotel workers in developing countries are too often trapped in poverty through poor working conditions. Working without a contract, going for months without pay or having to show up at the gate of the hotel in the morning to see if they are needed for the day are a few of the examples Tourism Concern found and which illustrate the perilous situation that people dependent on tourism may face (Tourism Concern n.d.).

Postmodernism

The concept of postmodernism has impacts on a variety of dimensions of tourism. Seen in part as a rejection of the more traditional and dominant ways of conceiving or operating, postmodernism has opened up new products, and tourists are demanding new and exciting experiences. Some have linked this to the shift from Fordism to Post-Fordism whereby there is a movement away from mass standardized packages. In order for developing countries to remain competitive they have to offer new products and open up more areas of their country to tourism. Demands for adventure tourism or ecotourism take people to areas previously not open to tourism. As indicated in Figure 8.1, there are resort areas opening up towards the interior of the island. Demand for heritage tourism is increasing as well, with tourists consuming culture and wanting to be educated. As noted in Chapter 4, heritage and cultural tourism raise issues of commodification and issues of access and control of heritage monuments. From a development perspective, these new types of products may well enhance the opportunities for locals to participate in the tourism industry. Any initiative that takes the tourists out from behind the resort walls provides the opportunity for local entrepreneurial activity. Again, the development dilemma appears, as opportunities may also come with associated costs (economic, environmental, social and political).

With the shift towards changing markets, there are pressures upon existing resort complexes to alter their factors of production. Hotels, for example, need to provide more varied activities, which means more equipment. Tour operators who are bringing in the tourists are competing in a very price-sensitive environment. They may place demands on resorts to offer more activities or products but they do not want to pass these costs on to the consumer. Therefore, the destination resort may have to absorb the new costs in meeting the higher demands of tour operators and tourists.

Consumption

The consumption of tourism as explored in Chapter 6 has a number of important implications for the extent to which tourism can play a meaningful role in the development process. Tourists were identified in Figure 4.1 as one of the agents of tourism development. How many tourists come, what they demand, how they behave, and specifically how much money they spend are all important factors in the development process. It is important to look at the markets that are attracted to developing countries. Often in a discussion of tourism the focus is on tourists from the developed countries who visit developing countries. What should not be forgotten is the role that domestic tourists can play in the development process. They may, for example, be more willing to use local forms of transportation, eat in local restaurants and stay in locally run accommodation. If one considers a country such as China and the potential in terms of numbers of future domestic tourists, this may well be a more important market than the international market. A news release from the World Tourism Organization from 14 June 2006 highlights the findings from a study of outbound tourism in Asia (WTO 2006b). The findings of this study reveal the importance not only of the domestic market but also the regional market. The study shows that 78 per cent of all international tourists come from another Asian country. The Chinese made 31 million international trips in 2005 and 91 per cent chose an Asian destination. For this market, group travel is losing popularity and independent travel is gaining. The top destinations for people from Hong Kong include Thailand, Japan and Singapore while the top five destinations of travellers from Thailand are Malaysia, Singapore, China, Hong Kong and Japan. A final example is Indian travellers, who made 6.1 million trips abroad in 2004 and they like to visit several countries during each trip (WTO 2006b). This study reveals the importance for developing countries to also target the more affluent sectors of the population in nearby developing countries.

The consumption of tourism is also linked to the discussion of postmodernism in the previous section. New demands for new products will not only create opportunities but also associated costs. As tourists become more experienced their demand profiles can change. There is an argument in the literature that tourists are becoming more aware of the impacts of travel and are more environmentally friendly. Figure 6.3 offers a code of ethics; however, as indicated in this chapter, there are debates about the effectiveness of such codes. While new forms of travel tend to get the most notice in the press and are frequently studied by academics, it is important not to forget the traditional mass tourism product, which has been the mainstay of tourism and will likely remain that way. There may be different labels put on it through marketing or there may be a perceived increase in the range of choice associated with these trips; however, mass tourism still represents the 'workhorse' of the tourism industry. It is also important to note that not all newer and smaller scale forms of tourism are sustainable and in fact it is some of the larger tourism corporations that are taking the lead in developing environmental initiatives. The challenge becomes to make all tourism more sustainable so that these will be further benefits to the host destination.

The final issue to explore in terms of consumption is what Potter (1995) terms 'convergence'. He argues that there is an increasing convergence on Western models and patterns of consumption. Economic dependency dovetails with cultural and psychological dependency (Potter 1995). In tourism a great deal has been written about the demonstration effect (see Chapter 6) and the impact that this has on local patterns of behaviour and consumption. Festivals and souvenirs have been changed to accommodate tourists. Consumption of some products will increase and other traditional products may fall away. These types of trade-offs are identified in Figure 8.1 and are discussed later in this chapter.

Sustainable development

Sustainable development was explored in detail in Chapter 2 and in this chapter a call was put forward in terms of the 'imperative of sustainable development'. Although the term is highly debated, is arguably a Western-based concept being imposed on developing countries and there are conflicts over how it should or can be implemented and measured, the concept continues to receive attention as a driving framework for tourism. The definition of development used in this book highlights the fact that improvements need to proceed within environmental limits. It is worth noting here that sustainability requires a holistic approach, covering not only environmental but also social, cultural, political and economic concerns. In structuring discussion to answer the question as to what it will take to move our civilization in the direction of long-term sustainable development, Wright (2008) focuses on two sets of themes. Strategic themes include sustainability, stewardship (the ethic that guides actions for the benefit of the natural world and other people) and science, and these are identified as concepts or ideas that can guide societies towards a sustainable future. Integrative themes include ecosystem capital (goods and services provided by ecosystems), policy and politics and globalization, and these three categories deal with the current status of interactions between natural systems and human societies. Tourism as a development tool interacts with all of these themes and steps are being taken to persuade businesses, tourists and destinations to be more sustainable. There are calls for more local participation in tourism planning and the operation of the industry (Chapters 4 and 5); however, as Moworth and Munt (1998) stress, sustainability needs to be understood within the context of power relations.

In an interesting study by Jamal *et al.* (2006), the authors state that while sustainable tourism and ecotourism are rooted in the notions of individual/societal and environmental well-being, they encountered conflicting values. They found that the actors and programmes relating to ecotourism have institutionalized a modernistic, commodifed paradigm. A question considered later in this chapter focuses on the ability to actually implement sustainability. If sustainability is adopted as a framework, what impacts will it have on the contribution towards the overall development of the destination? All forms of tourism need to be made more sustainable and Figure 8.1 explores the trade-offs that may occur in a destination. Having outlined some of the major influences on tourism in developing countries, the chapter now turns to an examination of the destination.

Tourism destination: form, function and response

The influences described in the previous section interact along with the tourism development process outlined in Chapter 4, resulting in some form of tourism. The island in the centre of Figure 8.1 represents that end-product and is illustrated here as having a highly developed resort area that is seeing not only beach development but also some form of satellite tourism development such as heritage tourism, ecotourism or nature-based tourism. While the diagram is of an island, many of the principles apply to other forms and scales of tourism. Markets are identified, and individuals, communities, companies, NGOs and governments respond to provide the product or the finances for the product. These development agents may come from within the destination or they may be external to it, which highlights the issues of control and power. Once a developing country decides to adopt tourism as a development option, governments need to decide whether to create and implement a tourism master plan or allow the industry to evolve with little interference from the government. If tourism plans are created, it is important to consider to what extent local people are involved in the planning process and whether they would actually be able to participate based on the constraints they face (e.g. economic, social, political, lack of knowledge).

In the centre of Figure 8.1 a series of elements are considered that reflect on how tourism is developed. All the agents of development encounter a policy, planning and regulatory environment (see Figure 4.1). Decisions will be taken by a number of different actors on what to build, who should finance it, who should operate it and what incentives, if any, should be provided to developers. For all key stakeholders there is a need to evaluate which tourism development options are the best and how the industry will compete in a global environment. The decisions made will vary from country to country over what forms of tourism to develop. There will also be a local response to tourism and, given the complex nature of communities, some groups will respond to the opportunities of tourism, some will passively accept the change, while others still may be more vocal or try to resist new developments.

Different types of tourism will assume different forms and functions, and how they are managed will also influence the degree to which they can contribute to development. Which ever resort or attraction is constructed (e.g. beach resort, village tourism, ecotourism lodge), it will act as a magnet for further development. In the case of beach resorts, it may even lead to the development of an urban tourism resort area. Over time, satellite tourist areas may open, as illustrated in Figure 8.1. Excursion roads to these attractions will themselves act as conduits for development as entrepreneurs try to sell their products to tourists at various rest stops or craft villages along the way. The overall impact spatially may well result in regional imbalances as tourism growth occurs around the tourism areas. The employment structure will change, especially if tourism develops in a major way. This will then lead to migration, and demands from immigrants for things such as housing, health care and schooling will need to be met. The challenge for any government is also how to redistribute the imbalances to ensure that those who do not work in the tourism sector and do not live near the tourism area will still be able to benefit from the industry. The final point raised at the centre of Figure 8.1 is the global–local interaction. These interactions are part of the tourism development trade-offs which are explored below.

Tourism development trade-offs

The development of tourism will invariably result in impacts in the destination as outlined in Chapter 7. As part of the tourism development dilemma, it will also typically result in some form of a trade-off. Broad examples of potential trade-offs are highlighted in Figure 8.1 and are meant as a framework for generating discussion rather than trying to present a comprehensive list. They are listed under the categories of economic, environmental, social/cultural and political. There can be trade-offs within and between these categories. The trade-offs may be the result of a conscious planning or policy decision (by a particular level of government, organization or business) or they may evolve over time as individuals, companies, organizations or governments respond to the evolving business environment. The trade-offs are not static but dynamic as the tourism industry evolves and responses are made (both positive and negative) to the industry. Trade-offs have important geographical dimensions including space and time (Adams 2005). In terms of space, tourism could be developed in a sustainable manner in one location and at the expense of the environment in another location. Adams (2005) notes the importance of both political and ecological boundaries. In terms of time, sustainability implies balancing resources for future generations with the needs of the present generation, and the future is hard to predict. Unintentional trade-offs can also occur as a planning decision in one direction or area may inadvertently cause change in another dimension of the destination (economic, environmental, social/cultural and political). Finally, the trade-offs will be regarded differently by different groups at different time periods. Some may be willing to support a trade-off while others will be against it.

Under the economic category, for example, tourism may bring economic growth for those able to participate in the tourism industry while at the same time it may result in economic leakages. If large-scale mass tourism is developed with multinational corporations, there will definitely be economic growth through increases in the numbers of tourists, development fees and taxes; however, if the multinational imports supplies and management personnel while repatriating profits, there will be a reduced multiplier effect. This may be a trade-off worth making if the developing country is not able to effectively enter the global tourism market without outside help. Liu and Wall (2006) argue that many and perhaps most communities in the developing world need an outside catalyst to stimulate interest in tourism and external expertise to take full advantage of their potential opportunities. Page and Connell (2006: 348) outline the economic costs and benefits of tourism. The economic benefits include balance of payments, income and employment while the economic costs include inflation, opportunity costs, dependency, seasonality and leakage. Income and employment can also be a cost if the better paid positions do not go to locals and if the income generated does not go to those who need it most (i.e. leaked out of the economy). Trade-offs under social/cultural could see a community-based tourism project promote empowerment in a local community; however, it may also lead to the commodification of their culture as modified souvenirs are sold to tourists that best suit their needs.

Trade-offs can also occur between a number of the main categories listed in Figure 8.1. Economic growth may be pursued at the cost of the environment and the loss of local control. Alternatively, ecotourism may be developed as a sustainable form of tourism; however, due to the restrictions in visitor numbers typically required, it may come at the cost of economic growth. It could be argued that enforcing strict and costly environmental policies may reduce the overall economic impact of tourism. An example of a social/cultural and economic trade-off could be accepting or tolerating more tourists for financial gain; yet a traditional sacred ceremony can become secular entertainment for tourists (see Plate 8.2). These are just a few of the multitude of trade-offs that can occur.

The question as to who decides on the nature of the trade-offs can largely be answered by examining the power relations that exist surrounding the tourism development process. At the national level the state may set out the regulatory environment which will form the basis for a particular type of trade-off (e.g.

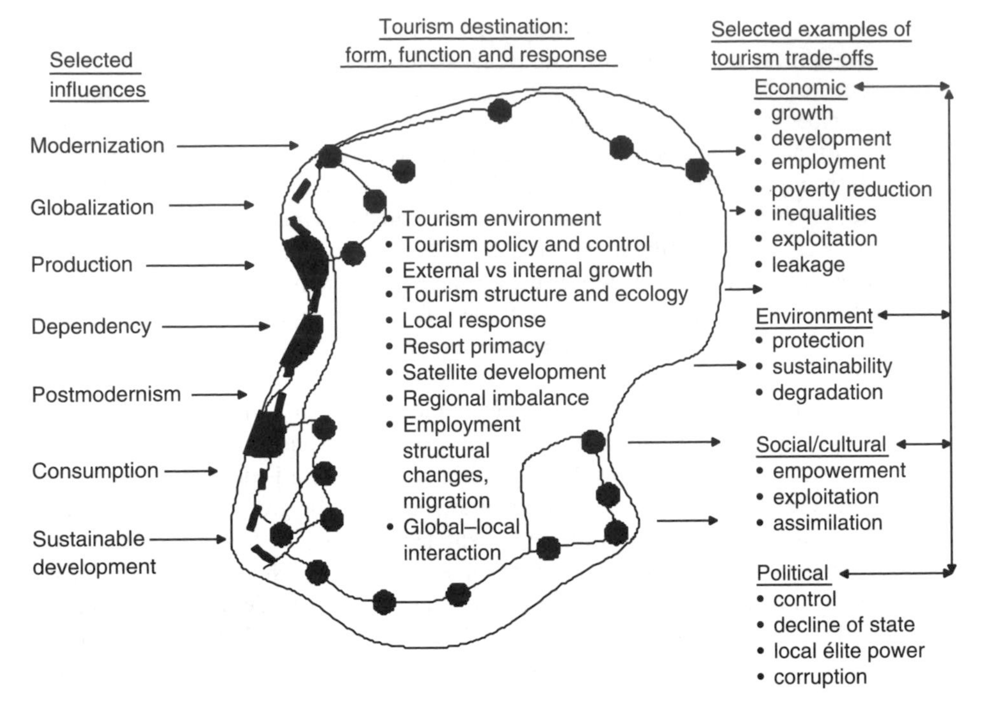

Figure 8.1 ***The tourism development dilemma framework.***

Source: After Potter (1995).

pursuing a policy of limiting all-inclusive resorts so that tourists will spend more money outside the resort – the trade-off being tour operators may be scared away) or conversely a powerful multinational may be able to dictate terms to local governments setting up a different type of trade-off (e.g. demand to be able to import significant quantities of food – the trade-off being that there may be more tourists as a result of the multinational; however, there will also be significant leakages). Power also rests with international funding agencies, which could dictate that a certain type of tourism development is built, or with the international money market, which can influence exchange rates and thereby dictate whether a particular country is attractive to invest in.

It is important to recognize that the tourism industry is a very dynamic industry, the interactions within and between the categories happen simultaneously, and there is not always an evident simple cause-and-effect relationship between the interactions in tourism. The best-laid plans may not always work and things may change in the future. There are a multiple of forces and actors playing out at the same time, as illustrated in Figure 8.1, and tourism will certainly not be the only industry in the destination, although it could be the largest. Within the framework of sustainability the strategy will hopefully minimize the level of negative trade-offs. One final trade-off considered here is that of opportunity cost. The decision to put money into tourism results in that same money not being available for

Plate 8.2 ***Indonesia, Kuta Beach Bali: Traditional Balinese ceremony on the beach continues in the presence of tourists.***

investment in another area of the economy; an area which could generate more development than tourism, depending on the situation.

Difficulties in implementing the ideals of sustainability

The trade-offs in tourism listed in Figure 8.1 often imply that one sector or element may benefit at the expense of another. Decision making means that competing needs and values must be considered so that the best conclusion can be reached in the numbing complexity of demands and circumstances (Wright 2008). Wright (2008) refers to sustainable development as an ideal, a goal towards which all human societies need to be moving, even if it has not been achieved completely anywhere. Sustainable solutions come at the intersection of the concerns of sociologists (socially desirable), economists (economically feasible) and ecologists (ecologically feasible) (Wright 2008). The idea of sustainable solutions raises the question of whether or not it is possible to move beyond the trade-offs in Figure 8.1 so that all three areas (environment, society and economy) benefit. In the case of tourism, for example, without an attractive environment tourists will not come, and no economic benefits will occur for the destination. If the locals are not able to participate in the planning, implementation and benefits of tourism, and are only the object of commodification, then there may be resentment of the locals towards the tourists. While the end goal in this hypothetical example may be to move in the direction of ensuring that all three areas (environment, society and economy) benefit, it is important to recognize that tourism is a highly competitive global industry, and corporations, governments, various organizations and individuals all have values and ideologies guiding behaviour and outcomes. Wright (2008), for example, states that when governments linked to anti-environmental special interests are in power, the environment, as well as sustainability, can take a beating. For now, as Adams (2005: 425) points out, the 'pursuit of sustainability demands choices about the distribution of costs and benefits in space and time'. The challenge then becomes how to make all forms of tourism more sustainable.

The weaknesses of sustainable development have been explored in Chapter 2 and a set of key questions is presented in Table 8.1. Challenges in implementation also relate to the policy process. Developing countries face global environmental agendas such as sustainable development as well as their own unique environmental problems and development needs (Newell 2005). An examination of environmental policy in developing countries illustrates both differences and common challenges in terms of reconciling development needs with long-term environmental goals (Newell 2005). Countries have responded differently to the challenge of reconciling these issues as a result of variations in political systems, the nature of their economies and the level of civil engagement. Various countries in the developing world have differing abilities to implement policies. Newell (2005) cites the example of India, which has some of the most impressive environmental legislation in the world; however, the lack of training resources and corruption of local pollution officials all act as constraints to implementation. In other situations where countries have a strong economic and developmental interest to ensure compliance with policies, extra steps are taken, such as in Kenya. Due to the significance of revenues from wildlife safaris in the country, officials have gone to controversial lengths in tackling the problem of illegal poaching of rhinos and elephants for their ivory as well as 'banning tribal groups from culling animals for food even on their own ancestral land' (Newell 2005: 228). Part of sustainable development is local involvement, and while countries such as India and Mexico have strong traditions in having active civil society engagement, these options are limited in places such as China or Singapore (Newell 2005). Newell (2005) also notes the differences to which there has been an overall 'greening' of industries, and the drivers of corporate environmental responsibility such as government incentives, consumer and investor pressure, and civil society watchdogs are currently undeveloped in many parts of the developing world. While there are often vast differences between countries, there have been similar problems in enforcement at the national level. Similar enforcement problems relate to constraints arising from economic relationships of trade, aid and debt, and conflicts between often Northern-determined environmental priorities and other issues which appear to be more pressing at local levels (Newell 2005).

Sustainable development is increasingly being linked to poverty alleviation and in this book the connections have been made to the Millennium Development goals. In another example of the challenges of implementation, Hawkins and Mann (2007: 360) comment that:

> though the Millennium Development Goal imperative is driving individual country development strategies, these indicators are often not relevant to the political agendas of individual government leaders or ministers. The Minister of Tourism in any given country now is being asked to deliver a range of outcomes, including economic growth that can be empirically related to a poverty reduction strategy.

They go on to refer to Ravallion (2004), who comments that economic growth does not necessarily lead to poverty reduction.

The above examples speak to the difficulties in implementing goals or ideas versus the realities in planning and managing tourism in developing countries. Wright (2008) argues, however, that there is increasing recognition from a wide variety of groups and individuals in different settings that 'business as usual' is not sustainable and new ways of thinking and conducting business will be required. The challenges in changing the way businesses, governments, organizations and individuals both think and act are very real, yet there are a number of initiatives, as outlined in Chapters 2 and 5, that are taking tourism in a more sustainable direction as well as linking tourism to poverty reduction.

Conclusion

Development is a highly contested term with a variety of different perspectives both in theory and practice. Development theory has changed as various paradigms have risen and fallen (though not completely disappearing), influencing how development is practised. Notions of development have changed from being primarily a focus on economic growth to a more holistic notion of development incorporating economy, environment and society. Increasingly, development has been linked to sustainable development and poverty reduction, as illustrated in the Millennium Development goals. For developing countries, tourism continues to be one of the favoured development strategies and there have been efforts to make tourism more sustainable or to have it directly relate to poverty reduction. In considering using tourism as an agent of development it is also necessary to frame the analysis with the broader issues of development as they relate to power and control. In this age of globalization, there is a wide range of actors involved directly or indirectly in the tourism development process including international agencies, states, the private sector, citizens' groups, NGOs and tourists, to name a few. All of these actors have their own values and goals that influence their actions and they each have a certain amount of resources, technology, power and control which they may be able to use to influence the outcomes of development. Underhill (2006: 19) argues that the exercise of power in the international political economy takes place in a setting which is characterized by the complex interdependence among states and their societies and the economic structures at domestic and international levels. There are debates over the influence the free market, the state or civil society should have on the tourism development process. This book has examined the role tourism can play in the overall development process and, as Addison (2005: 219) suggests,

> getting development policy right has the potential to lift millions out of poverty and misery. Making the right policy choices is not just a technical matter. It requires careful political judgement on how to promote economic and social change in ways that stand the best chance of succeeding.

The decision to pursue tourism leads to the tourism development dilemma, which has been the focus of this book. Developing countries face significant challenges; tourism not only generates benefits, it also generates costs, and trade-offs are made. These trade-offs occur not only in the destination but also across national boundaries as well as across time. This chapter began by looking at the development imperative and the sustainability imperative. Sustainability has come to the forefront of the development debate and it is argued that all forms of tourism need to be more sustainable. It is presented in this chapter as an imperative involving not only environmental concerns but also social and economic considerations. Local communities need to have greater participation in the planning and development process. In the context of Figure 8.1, the aim of sustainability would be to minimize the negative trade-offs within the tourism development dilemma. Although there is a great deal of debate surrounding the concept, efforts on a number of different fronts are being made to try to make tourism more sustainable.

With the shift towards globalization there are concerns in the area of sustainability as multinational corporations seek out favourable destinations with good investment incentives and low production costs. The lower production costs may revolve around less stringent regulations on such areas as the environment or labour laws. The globalization debate in tourism has also raised issues of dependency and loss of control. There may be a reduction in the power of the state if a country wishing to enter the global tourism market is pressured into opening up its borders to multinationals. Globalization not only has economic, political and environmental consequences, it also has cultural implications for developing countries.

The government in a destination establishes the policy framework for tourism in its country as well as interacting with global corporations, and so it continues to play an important role. Many international agencies stress good governance. An open policy on tourism with numerous incentives will attract developers; however, caution needs to be taken when the priorities of a corporation do not match those of the state. These competing interests in tourism development can be illustrated through the continuum for advice on tourism planning developed by Burns (1999a). At one end of the continuum is 'Tourism First' where the industry is the focus of planning. At the other end is 'Development First' where planning is done according to national developmental needs. A stronger planning focus on 'Development First' may facilitate a more direct link to poverty reduction in the context of the Millennium Development goals. If a country adopts a policy to support large-scale projects with foreign investment, it also needs to make sure it supports small, locally controlled projects with things such as micro-credits where the informal sector may have a greater role to play. Communities respond to the opportunities of tourism and they need to be heard to effectively involve more people in the benefits of tourism. A number of interesting programmes, such as fair trade in tourism and pro-poor tourism, have been raised to specifically target poverty; however, there are significant challenges that need to be overcome. It is also important that development is not planned in isolation from the rest of the economy as it needs to be integrated into multiple sectors so that there is the potential for more benefits to accrue.

While tourism development is often centred on governments, planners, corporations, communities and NGOs, it is important not to forget the role of the tourist in the development process. Decisions are made as to what markets are pursued. Different types of tourists will exhibit different types of behaviour and will have different spending patterns. Some countries have opted for large-scale mass tourism while others focus on smaller numbers of higher paying tourists, while others still have pursued a more mixed approach attracting a diversity of markets. How a country chooses to market and brand itself will influence the overall type of development along with the types of tourists who visit the destination.

All tourism developments have impacts, and different communities and environments may be better able to handle these impacts than others. These impacts need not always be negative as communities and individuals respond to the opportunities presented by tourism (Wall and Mathieson 2006). For many developing countries there is limited choice in terms of development options and they may, in fact, have a real competitive advantage on a global scale to offer tourism services. In fact, as Lickorish (1991: 162) argues, tourism may prove more 'cost effective in its use of scarce foreign investment for creating additional prosperity than alternatives in primary or manufacturing industry'.

Tourism is but one agent of development being actively pursued by many countries, and what seems to work in one developing destination may not work in another. The relationship between tourism and development is multifaceted as theories, values, actors, power, strategies, politics, policies, plans, communities and environments all interact in a dynamic global system. Exploring the interaction of these concepts and issues will help us better understand the nature of the tourism development dilemma for developing countries.

Discussion questions

1. **To what extent does tourism promote development and increase the standard of living for the local population?**
2. **Does sustainability represent a way forward to guide tourism towards promoting development or does it act as a barrier to development?**
3. **How can development be measured with respect to tourism?**
4. **What processes or goals should guide tourism development over the next 25 years?**

Further reading

Harrison, D. (ed.) (2001) *Tourism and the Less Developed World: Issues and Case Studies*, New York: CABI. [This book has an introductory section on issues relating to tourism development in developing countries before moving on to 13 chapters dealing with a variety of topics/case studies.]

Huybers, T. (ed.) (2007) *Tourism in Developing Countries*, Cheltenham: Edward Elgar. [This edited volume contains 34 articles dating from 1974 to 2004 that explore a range of issues for tourism in developing countries. The book investigates the positive and negative aspects (economic, social-cultural and environmental) associated with tourism development.]

Johnston, R., Taylor, P. and Watts, M. (eds) (2005) *Geographies of Global Change Remapping the World*, Oxford: Blackwell. [This book explores the nature of global change in 28 chapters. The authors examine economic, political, social, cultural and ecological dimensions of change.]

Websites

The webpage for the UK's Department for International Development (DFID). The work of the DFID is highlighted and the 2000 White Paper on *Eliminating World Poverty – Making Globalisation Work for the Poor* is available through the site: www.dfid.gov.uk.

The web page for the Canadian International Development Agency (CIDA). The site has links to the various programmes it has supported. For example, CIDA provides assistance to the Organisation of Eastern Caribbean States (OECS) in the areas of trade negotiations, assisting small and medium-sized enterprises as they adjust to global markets, as well safeguarding the tourism industry though protecting the natural environment: www.acdi-cida.gc.ca/index.htm.

References

Accor (2006) Accor, www.accor.com, accessed 8 August 2006.

Adams, W. (2005) Sustainable development?, in R. Johnston, P. Taylor and M. Watts (eds) *Geographies of Global Change Remapping the World*, Oxford: Blackwell, pp. 412–426.

Addison, T. (2005) Development, in P. Burnell and V. Randall (eds) *Politics in the Developing World*, Oxford: Oxford University Press, pp. 205–230.

Aguiló, E., Alegre, J. and Sard, M. (2005) The persistence of the *sun and sand* tourism model, *Tourism Management* 26: 219–231.

Akis, S., Peristianis, N. and Warner, J. (1996) Resident attitudes to tourism development: the case of Cyprus, *Tourism Management* 17(7): 481–494.

Allen, J. (1995) Global worlds, in J. Allen and D. Massey (eds) *Geographical Worlds*, Oxford: Oxford University Press, pp 105–144.

Andersen, V. (1991) *Alternative Economic Indicators*, London: Routledge.

Andreassen, B. and Marks, S. (2006) *Development as a Human Right*, London: Harvard School of Public Health.

Ap, J. and Crompton, J. (1993) Residents' strategies for responding to tourism impacts, *Journal of Travel Research* 31(3): 47–50.

Arai, S. (1996) Benefits of citizen participation in a healthy community initiative: linking community development and empowerment, *Journal of Applied Recreation Research* 21: 25–44.

Archabald, K. and Naughton-Treves, L. (2001) Tourism revenue-sharing around national parks in Western Uganda: early efforts to identify and reward communities, *Environmental Conservation* 28(2): 135–140.

Arnstein, S. (1969) A ladder of citizen participation, *American Institute of Planners Journal* July: 216–224.

Ashley, C., Boyd, C. and Goodwin, H. (2000) Putting poverty at the heart of the tourism agenda, *Natural Resource Perspectives* No. 51, London: Overseas Development Institute.

Ashley, C., Goodwin, H., McNab, D., Scott, M. and Chaves, L. (2006) *Making Tourism Count for the Local Economy in the Caribbean*: www.propoortourism.org.uk/caribbean/caribbean-brief-whole.pdf.

Associated Press (2006) New law worries India's young workers, *The Standard*, 11 October: B8.

Atlantica Hotels (2006) Atlantica Hotels, www.atlanticahotels.com.br, accessed 27 September 2006.

Azarya, V. (2004) Globalisation and international tourism in developing countries: marginality as a commercial commodity, *Current Sociology* 52(6): 949–967.

Bah, A. and Goodwin, H. (2003) *Improving Access to the Informal Sector to Tourism in The Gambia*, PPT Working Paper No. 15, London: Pro-poor Tourism Partnership.

Baker, S. (2006) *Sustainable Development*, London: Routledge.

Barkin, D. and Bouchez, C. (2002) NGO–Community collaboration for ecotourism: a strategy for sustainable regional development, *Current Issues in Tourism* 5(3 and 4): 245–253.

Barratt Brown, M. (1993) *Fair Trade: Reform and Realities in the International Trading System*, London: Zed Books.

Bastin, R. (1984) Small island tourism: development or dependency?, *Development Policy Review* 2(1): 79–90.

Baud-Bovy, M. and Lawson, F. (1998) *Tourism and Recreation Handbook of Planning and Design*, Oxford: Architectural Press.

Bauer, T. and McKercher, B. (2003) *Sex and Tourism: Journeys of Romance, Love and Lust*, New York: Haworth Hospitality Press.

Becken, S. and Simmons, D. (2005) Tourism, fossil fuel consumption and the impact on the global climate, in C. M. Hall and J. Higham (eds) *Tourism, Recreation and Climate Change*, Clevedon: Channel View Publications, pp. 192–206.

Bell, C. and Newby, H. (1976) Communion, communalism, class and community action: the sources of new urban politics, in D. Herbert and R. Johnston (eds) *Social Areas in Cities Volume 2*, Chichester: Wiley.

Berno, T. and Bricker, K. (2001) Sustainable tourism development: the long road from theory to practice, *International Journal of Economic Development* 3(3): 1–18.

Bianchi, R. (2002) Towards a new political economy of global tourism, in R. Sharpley and D. J. Telfer (eds) *Tourism and Development: Concepts and Issues*, Clevedon: Channel View Publications, pp. 265–299.

Bigano, A., Hamilton, J., Lau, M, Tol, R. and Zhou, Y. (2004) *A Global Database of Domestic and International Tourist Numbers at National and Subnational Level*, www.uni-hamburg.de/Wiss/FB/15/Sustainability/tourismdata.pdf.

Bintan Resorts (2006) Investing in Bintan Resorts, www.bintan-resort.com, accessed 11 August 2006.

Bjorklund, E. M. and Philbrick, A. K. (1972) *Spatial configurations of mental processes*, unpublished paper, Department of Geography, University of Western Ontario, London, Ontario.

Blackstock, K. (2005) A critical look at community based tourism, *Community Development Journal* 40(1): 39–49.

Boissevan, J. (1996) Introduction, in J. Boissevan (ed.) *Coping with Tourists, European Reactions to Mass Tourism*, Oxford: Berghan Books, pp.1–26.

Bookman M. (2006) *Tourists, Migrants and Refugees, Population Movements in Third World Development*, London: Lynne Rienner.

Borger, J. (2006) Half of global car exhaust produced by US vehicles, *Guardian*, 29 June.

Boulding, K. (1992) The economics of the coming spaceship earth, in A. Markandya and J. Richardson (eds) *The Earthscan Reader in Environmental Economics*, London: Earthscan, pp. 27–35.

Bramwell, B. and Lane, B. (1993) Sustainable tourism: an evolving global approach, *Journal of Sustainable Tourism* 1(1): 1–5.

Bramwell, B. and Lane, B. (2000) *Tourism Collaboration and Partnership Politics, Practice and Sustainability*, Clevedon Hall: Channel View Publications.

Brinkerhoff, D. W. and Ingle, M. D. (1989) Integrating blueprint and process: a structured flexibility approach to development management, *Public Administration And Development* 9: 487–503.

Britton, S. (1982) The political economy of tourism in the Third World, *Annals of Tourism Research* 9(3): 331–58.

Britton, S. (1991) Tourism, capital and place: towards a critical geography of tourism, *Environment and Planning D: Society and Space* 9: 451–478.

Brohman, J. (1996a) New directions for tourism in Third World development, *Annals of Tourism Research* 23(1): 48–70.

Brohman, J. (1996b) *Popular Development: Rethinking the Theory and Practice of Development*, Oxford: Blackwell.

Brown, D. (2006) Gap year projects slammed as out-dated, Eturbonews, retrieved 18 August from www.travelwirenews.com/cgi-script/csArticles/articles/00009432-p.htm.

Brown, F. (1998) *Tourism Reassessed: Blight or Blessing?*, Oxford: Butterworth-Heinemann.

Brown, T. (1999) Antecedents of culturally significant tourist behaviour, *Annals of Tourism Research* 26(3): 676–700.

Brunet, S., Bauer, J., De Lacy, T. and Tshering, K. (2001) Tourism development in Bhutan: tensions between tradition and modernity, *Journal of Sustainable Tourism* 9(3): 243–263.

Buckley, R. (2003) *Case Studies in Ecotourism*, Wallingford: CABI Publishing.

Buhalis, D. (2000) Marketing the competitive destination of the future, *Tourism Management* 21(1): 97–116.

Buhalis, D. and Ujma, D. (2006) Intermediaries: travel agencies and tour operators, in C. Costa ad D. Buhalis (eds) *Tourism Management Dynamics, Trends, Management and Tools*, London: Elsevier Butterworth Heinemann, pp. 172–180.

Burns, P. (1999a) Paradoxes in planning, tourism elitism or brutalism, *Annals of Tourism Research* 26(2): 329–348.

Burns, P. (1999b) Editorial – Tourism NGOs, *Tourism Recreation Research* 24(2): 3–6.

Burns, P. and Holden, A. (1995) *Tourism: A New Perspective*, Hemel Hempstead: Prentice Hall International.

Butcher, J. (2002) *The Moralisation of Tourism: Sun, Sand . . . and Saving the World?*, London: Routledge.

Butler, R. W. (1975) Tourism as an agent of social change, in F. Helleiner (ed.) *Tourism as a Factor in National and Regional Development*, Occasional Paper No. 4, Department of Geography, Trent University, Peterborough, Ontario, pp. 85–90.

Butler, R. W. (1980) The concept of the tourist area cycle of evolution: implications for managers of resources, *Canadian Geographer* 24(1): 5–12.

Butler, R. W. (1990) Alternative tourism: pious hope or Trojan horse?, *Journal of Travel Research* 28(3): 40–45.

Butler, R. W. (1993) Tourism – an evolutionary perspective, in J. Nelson, R. Butler and G. Wall (eds) *Tourism and Sustainable Development: Monitoring, Planning, Managing*, University of Waterloo: Department of Geography, pp. 27–43.

Butler, R. W. (1994) Alternative tourism: the thin edge of the wedge, in V. Smith and W. Eadington (eds) *Tourism Alternatives, Potential and Problems in the Development of Tourism*, Chichester: John Wiley & Sons, pp. 31–46.

Campbell, F. (1999) Whispers and waste, *Our Planet* 10(3) ('Small Islands'), www.ourplanet.com.

Campbell, L. and Smith, C. (2006) What makes them pay? Values of volunteer tourists working for sea turtle conservation, *Environmental Management* 38(1): 84–98.

Carson, R. (1962) *Silent Spring*, Boston, MA: Houghton Mifflin.

Cater, E. (1993) Ecotourism in the Third World: problems for sustainable tourism development, *Tourism Management* 14(2): 85–93.
Cater, E. (1994) Ecotourism in the Third World – problems and prospects for sustainability, in E. Cater and G. Lowman (eds) *Ecotourism A Sustainable Option?*, Chichester: John Wiley & Sons, pp. 69–86.
Cater, E. (2004) Ecotourism: theory and practice, in A. Lew, C. M. Hall and A. Williams (eds) *A Companion to Tourism*, Oxford: Blackwell, pp. 484–497.
Cerviño, J. and Cubillo, M. (2005) Hotel and tourism development in Cuba, *Cornell Hotel and Restaurant Administration Quarterly* 46(2): 223–246.
Chambers, D. and Airey, D. (2001) Tourism policy in Jamaica: A Tale of Two Governments, *Current Issues in Tourism* 4(2–4): 94–120.
Chambers, R. (1997) *Whose Reality Counts? Putting the First Last*, London: Intermediate Technology Publications.
Chandler, P. (1999) Fair trade in tourism – the independent tour operator, in Tourism Concern, Voluntary Service Overseas, University of North London, *Achieving Fairly Traded Tourism Conference*, Twickenham: London and Association of Independent Tour Operators.
Choi, H. and Sirakaya, E. (2006) Sustainability indicators for managing community tourism, *Tourism Management* 27: 1274–1289.
Chok, S., Macbeth, J. and Warren, C. (2007) Tourism as a tool for poverty alleviation: a critical analysis of 'pro-poor tourism' and implications for sustainability, *Current Issues in Tourism* 10(2 and 3): 144–165.
Clancy, M. (1999) Tourism and development: evidence from Mexico, *Annals of Tourism Research* 26(1): 1–20.
Cleverdon, R. (2001) Introduction: Fair trade in tourism – applications and experience, *International Journal of Tourism Research* 3: 347–349.
Cleverdon, R. and Kalisch, A. (2000) Fair trade in tourism, *International Journal of Tourism Research* 2: 171–187.
Cohen, E. (1972) Towards a sociology of international tourism, *Social Research* 39(1): 64–82.
Cohen, E. (1979) A phenomenonology of tourist experiences, *Sociology* 13: 179–201.
Cohen, E. (1982) Marginal paradises: bungalow tourism on the islands of southern Thailand, *Annals of Tourism Research* 9(2): 189–228.
Colantonio, A. and Potter, R. (2006) The rise of urban tourism in Havana since 1989, *Geography* 91(1): 23–33.
Coles, T. and Church, A. (2007) Tourism, politics and the forgotten entanglements of power, in A. Church and T. Coles (eds) *Tourism, Power and Space*, London: Routledge, pp.1–42.
Cooper, C., Fletcher, J., Fyall, A., Gilbert, D. and Wanhill, S. (2005) *Tourism: Principles and Practice* (3rd edn), Harlow: Pearson Education.
Cooper, M. (2003) The real Cancún: behind globalization's glitz, *The Nation*, 22 September.
Cooperative Bank (2005) *The Ethical Consumerism Report 2005*, Manchester: The Cooperative Bank.
Costa, C. (2006) Tourism planning, development and territory, in D. Buhalis and C. Costa (eds) *Tourism Management Dynamics, Trends, Management and Tools*, London: Elsevier Butterworth Heinemann, pp. 236–243.
Costa, C. and Buhalis, D. (2006) Introduction, in C. Costa and D. Buhalis (eds) *Tourism Management Dynamics, Trends, Management and Tools*, London: Elsevier Butterworth Heinemann, pp. 1–5.
Countryside Commission (1995) *Sustainable Rural Tourism: Opportunities for Local Action*, CCP 483, Cheltenham: Countryside Commission.

Cowe, R. and Williams, S. (2000) *Who is the Ethical Consumer?*, Manchester: The Cooperative Bank.

Cowen, M. and Shenton, R. (1996) *Doctrines of Development*, London: Routledge.

Croall, J. (1995) *Preserve or Destroy: Tourism and the Environment*, London: Calouste Gulbenkian Foundation.

Crompton, J. (1979) Motivations for pleasure vacation, *Annals of Tourism Research* 6(4): 408–424.

Cronin, L. (1990) A strategy for tourism and sustainable developments, *World Leisure and Recreation* 32(3): 12–18.

Cukier, J. (2002) Tourism employment issues in developing countries: examples from Indonesia, in R. Sharpley and D. J. Telfer (eds) *Tourism and Development Concepts and Issues*, Clevedon: Channel View Publications, pp. 165–201.

Cukier, J. and Wall, G. (1994) Informal tourism employment: vendors in Bali, Indonesia, *Tourism Management* 15(6): 464–467.

Daher, R. (2005) Urban regeneration/heritage tourism endeavours: the case of Salt, Jordan, local actors, international donors and the state, *International Journal of Heritage Studies* 11(4): 289–308.

D'Amico, B. (2005) *A Touch of Africa, Part II on to the Amazon*, Bloomington: Authorhouse.

Dann, G. (1981) Tourist motivation: an appraisal, *Annals of Tourism Research* 8(2): 187–219.

Dasgupta, P. and Weale, M. (1992) On measuring the quality of life, *World Development* 20(1): 119–131.

Davis, H. D. and Simmons, J.A. (1982) World Bank experience with tourism projects, *Tourism Management* 3(4): 212–217.

de Araujo, L. and Bramwell, B. (1999) Stakeholder assessment and collaborative tourism planning: the case of Brazil's Costa Dourada Project, *Journal of Sustainable Tourism* 7(3 and 4): 356–378.

de Holan, P. and Phillips, N. (1997) Sun, sand and hard currency: tourism in Cuba, *Annals of Tourism Research* 24(4): 777–795.

de Kadt, E. (1979) *Tourism: Passport to Development?*, New York: Oxford University Press.

de Rivero, O. (2001) *The Myth of Development: Non-viable Economies of the 21st Century*, London: Zed Books.

Desai. V. and Potter, R. (2002) *The Companion to Development Studies*, London: Arnold.

Diamantis, D. (1999) Green strategies for tourism worldwide, *Travel & Tourism Analyst* 4: 9–112.

Diamantis, D. (2004) Ecotourism management an overview, in D. Diamantis (ed.) *Ecotourism: Management and Assessment*, London: Thomson, pp. 1–26.

Diamond, J. (1997) Tourism's role in economic development: the case re-examined, *Economic Development and Cultural Change* 25(3): 539–553.

Din, K. (1982) Tourism in Malaysia: competing needs in a plural society, *Annals of Tourism Research* 16(4): 453–480.

Dogan, H. (1989) Forms of adjustment: sociocultural impacts of tourism, *Annals of Tourism Research* 16(2): 216–236.

Dollar, D. (2005) Globalization, poverty and inequality, in M. Weinstein (ed.) *Globalization What's New?*, New York: Columbia University Press, pp. 96–128.

Dorji, T. (2001) Sustainability of tourism in Bhutan, *Journal of Bhutan Studies* 3(1): 84–104.

Dowling, R. (1992) Tourism and environmental integration; the journey from idealism to realism, in C. Cooper and A. Lockwood (eds) *Progress in Tourism, Recreation and Hospitality Management*, Vol. 4, London: Bellhaven Press, pp. 33–46.

Doxey, G. (1975) A causation theory of visitor–resident irritants: methodology and research inferences, *Proceedings of the Travel Research Association*, 6th Annual Conference, San Diego, pp. 195–198.

Doxey, G. (1976) When enough's enough: the natives are restless in Old Niagara, *Heritage Canada* 2(2): 26–27.

Dresner, S. (2002) *The Principles of Sustainability*, London: Earthscan Publications.

Duffy, R. (2000) Shadow players: ecotourism development, corruption and state politics in Belize, *Third World Quarterly* 21(3): 549–565.

Duffy, R. (2002) *A Trip Too Far*, London: Earthscan.

Duffy, R. (2006) The politics of ecotourism and the developing world, *Journal of Ecotourism* 5(1 and 2): 1–6.

Easterly, W. (2005) The rich have markets, the poor the bureaucrats, in M. Weinstein (ed.) *Globalization What's New?*, New York: Columbia University Press, pp. 170–195.

Eber, S. (1992) *Beyond The Green Horizon: Principles for Sustainable Tourism*, Godalming: WWF.

EC (1993) *Taking Account of Environment in Tourism Development*, DG XXIII Tourism Unit, Luxembourg: Commission of the European Communities.

Echtner, C. and Prasad, P. (2003) The context of Third World tourism marketing, *Annals of Tourism Research* 30(3): 660–682.

Elegant, S. (2006) Rooms to grow, service can't be knocked off, one reason Western hotels are betting big on China, *Time* 168(21): A13–14.

Elliott, J. (1997) *Tourism and Public Policy*, London: Routledge.

ETB (1991) *The Green Light: A Guide to Sustainable Tourism*, London: English Tourist Board.

Evans, G. (2005) Mundo Maya: from Cancún to city of culture. World heritage in post-colonial Mesoamerica, in D. Harrison and M. Hitchcock (eds) *The Politics of World Heritage Negotiating Tourism and Conservation*, Clevedon: Channel View Publications, pp. 35–49.

Evans, G. and Cleverdon, R. (2000) Fair trade in tourism – community development or marketing tool, in D. Hall and G. Richards (eds) *Tourism and Sustainable Community Development*, London: Routledge, pp. 137–153.

FAO (2006) *The State of Food Insecurity in the World*, FAO United Nations, www.fao.org, accessed 1 November.

Farver, J. (1984) Tourism and employment in The Gambia, *Annals of Tourism Research* 11(2): 249–265.

Fennell, D. (1999) *Ecotourism: An Introduction*, London: Routledge.

Fennell, D. (2006) *Tourism Ethics*, Clevedon: Channel View Publications.

Fonatur (2006) About Fonatur, http://www.fonatur.gob.mx, accessed 21 September 2006.

Foo, J., McGuiggan, R. and Yiannakis, A. (2004) Roles tourists play: an Australian perspective, *Annals of Tourism Research* 31(2): 408–427.

Forsyth, T. (1995) Business attitudes to sustainable tourism: self-regulation in the UK outgoing tourism industry, *Journal of Sustainable Tourism* 3(4): 210–231.

Freeman, M. (2005) Human rights, in P. Burnell and V. Randall (eds) *Politics in the Developing World*, Oxford: Oxford University Press, pp. 237–252.

Galtung, J. (1986) Towards a new economics: on the theory and practice of self-reliance, in P. Ekins (ed.) *The Living Economy: A New Economy in the Making*, London: Routledge, pp. 97–109.

Garrod, B. and Fennell, D. (2004) An analysis of whalewatching codes of conduct, *Annals of Tourism Research* 31(2): 334–352.

Gartner, W. (2004) Factors affecting small firms in tourism: a Ghanaina perspective, in

R. Thomas (ed.) *Small Firms in Tourism: International Perspectives*, Oxford: Elsevier, pp. 35–70.

Getz, D. (1987) Tourism planning and research: traditions, models and futures, paper presented at the Australian Travel Research Workshop, Bunbury, Western Australia, 5–6 November.

Ghimire, K. (2001) The growth of national and regional tourism in developing countries: an overview, in K. Ghimire (ed.) *The Native Tourist: Mass Tourism within Developing Countries*, London: Earthscan, pp. 1–29.

Global Travel Industry News (2007) Asian low-cost carriers go international, Global Travel Industry News, www.trevelindustryreview.com/news/5423, accessed 16 May 2007.

Go, F. and Pine, R. (1995) *Globalisation Strategy in the Hotel Industry*, London: Routledge.

Godfrey, K. (1996) Towards sustainability? Tourism in the republic of Cyprus, in L. Harrison and W. Husbands (eds) *Practising Responsible Tourism: International Case Studies in Tourism Planning, Policy and Development*, Chichester: John Wiley & Sons, pp. 58–79.

Gold, J. (1980) *An Introduction to Behavioural Geography*, Oxford: Oxford University Press.

Goldsworthy, D. (1988) Thinking politically about development, *Development and Change* 19(3): 505–530.

Goodland, R. (1992) The case that the world has reached its limits, in R. Goodland, H. Daly, S. Serafy and B. von Droste (eds) *Environmentally Sustainable Economic Development: Building on Brundtland*, Paris: UNESCO, pp. 15–27.

Gössling, S. and Schultz, U. (2005) Tourism-related migration in Zanzibar, Tanzania, *Tourism Geographies* 7(1): 43–62.

Goudie, A. and Viles, H. (1997) *The Earth Transformed: An Introduction to Human Impacts on the Environment*, Oxford: Blackwell.

Goulet, D. (1992) Participation in development: new avenues, *World Development* 17(2): 165–178.

Graham, A. (2006) Transport and transit: air, land and sea, in C. Costa and D. Buhalis (eds) *Tourism Management Dynamics, Trends, Management and Tools*, London: Elsevier Butterworth Heinemann, pp. 181–190.

Gray, H. (1970) *International Travel – International Trade*, Lexington: DC Heath.

Gunn, C. with Var, T. (2002) *Tourism Planning Basics, Concepts, Cases* (4th edn), New York: Routledge.

Hall, C. M. (1994) *Tourism and Politics: Policy, Power and Place*, Chichester: John Wiley & Sons.

Hall, C. M. (2000) *Tourism Planning: Policies, Processes and Relationships*, Harlow: Prentice Hall.

Hall, C. M. (2002) Local initiatives for local regional development: the role of food, wine and tourism, in E. Arola, J. Kärkkäinen and M. Siltari (eds) *Tourism and Well Being: The 2nd Tourism Industry and Education Symposium May 16–18, Jyväsklä, Finland Symposium Proceedings*, Jyväsklä Polytechnic, pp. 47–63.

Hall, C. M. (2005) *Tourism Rethinking The Social Science of Mobility*, London: Pearson Prentice Hall.

Hall, C. M. (2007a) Tourism, governance and the (mis-) location of power, in A. Church and T. Coles (eds) *Tourism, Power and Space*, London: Routledge, pp. 247–268.

Hall, C. M. (2007b) Editorial, Pro-poor tourism: do tourism exchanges benefit primarily the countries of the south?, *Current Issues in Tourism* 10(2 and 3): 111–118.

Hall, C. M. and Higham, J. (eds) (2005) *Tourism, Recreation and Climate Change*, Clevedon: Channel View Publications.

Hall, C. M. and Jenkins, J. (1998) The policy dimensions of rural tourism and recreation, in R. Butler, C. M. Hall and J. Jenkins (eds) *Tourism and Recreation in Rural Areas*, Chichester: Wiley, pp. 19–42.

Hall, C. M. and Jenkins, J. (2004) Tourism and public policy, in A. Lew, C. M. Hall and A. M. Williams (eds) *A Companion to Tourism*, Oxford: Blackwell, pp. 525–540.

Hall, C. M. and Page, S. (2006) *The Geography of Tourism and Recreation* (3rd edn), London: Routledge.

Hall, D. (2000) Identity, community and sustainability prospects for rural tourism in Albania, in D. Hall and G. Richards (eds) *Tourism and Sustainable Community Development*, London: Routledge, pp. 48–59.

Hannam, K., Sheller, M. and Urry, J. (2006) Editorial: Mobilities, immobilities and moorings, *Mobilities* 1(1): 1–22.

Hard Rock Café (2006) Hard Rock Café advertisement *in easyJet inflight magazine*, July.

Hardin, G. (1968) The tragedy of the commons, *Science* 162: 1243–1248.

Harrigan, J. and Mosley, P. (1991) Evaluating the impact of World Bank structural adjustment lending, *Journal of Development Studies* 27(3): 63–94.

Harrison, D. (1988) *The Sociology of Modernisation and Development*, London: Routledge.

Harrison, D. and Price, M. (1996) Fragile environments, fragile communities? An introduction, in M. Price and V. Smith (eds) *People and Tourism in Fragile Environments*, Chichester: John Wiley & Sons, pp. 1–18.

Harrison, D. and Schipani, S. (2007) Lao tourism and poverty alleviation: community-based tourism and the private sector, *Current Issues in Tourism* 10(2 and 3): 194–230.

Harvey, D. (1989) The *Condition of Postmodernity*, Oxford: Blackwell.

Hashimoto, A. (2002) Tourism and sociocultural development issues, in R. Sharpley and D. J. Telfer (eds.) *Tourism and Development Concepts and Issues*, Clevedon: Channel View Publications, pp. 202–230.

Hashimoto, A. and Telfer, D. J. (2007) Geographic representations embedded within souvenirs in Niagara: the case of geographically displaced authenticity, *Tourism Geographies* 9(2): 191–217.

Hatton, M. (1999) *Community-based Tourism in the Asia-Pacific*, Toronto: Canadian Tourism Commission, Asia-Pacific Economic Cooperation and Canadian International Development Agency.

Hawkins, D. and Mann, S. (2007) The World Bank's role in tourism development, *Annals of Tourism Research* 34(2): 348–363.

Hazbun, W. (2004) Globalisation, reterritorialisation and the political economy of tourism development in the Middle East, *Geopolitics* 9(2): 310–341.

Hecho En GeoCuba (1997) Las Terrazas Complejo Turistico Tourist Map.

Held, D. (ed.) (2000) *A Globalizing World?: Culture, Economics, Policy*, London: Routledge.

Helleiner, E. (2006) Alternatives to neoliberalism? Towards a more heterogeneous global political economy, in R. Stubbs and G. Underhill (eds) *Political Economy and the Changing Global Order* (3rd edn), Don Mills, Ontario: Oxford University Press. pp. 77–87.

Henderson, J. (2003) The politics of tourism in Myanmar, *Current Issues in Tourism* 6(2): 97–118.

Henderson, J. (2006) Tourism in Dubai: overcoming barriers to destination development, *International Journal of Tourism Research* 8(2): 87–99.

Hettne, B. (1995) *Development Theory and the Three Worlds*, New York: Longman.

Hettne, B. (2002) Current trends and future options in development studies, in V. Desai and R. B. Potter (eds) *The Companion to Development Studies*, New York: Oxford University Press. pp. 7–12.

Hiernaux-Nicolas, D. (1999) Cancún bliss, in D. Judd and S. Fainstein (eds) *The Tourist City*, New Haven, CT: Yale University Press, pp. 124–139.

Hinch, T. and Butler, R. (1996) Indigenous tourism: a common ground for discussion, in R. Butler and T. Hinch (eds) *Tourism and Indigenous Peoples*, London: International Thomson Business Press, pp. 3–19.

Hitchcock, M. and Darma Putra, I. (2005) The Bali bombings: tourism crisis management and conflict avoidance, *Current Issues in Tourism* 8(1): 62–76.

Høivik, T. and Heiberg, T. (1980) Centre–periphery tourism and self-reliance, *International Social Science Journal* 32(1): 69–98.

Holden, A. (2000) *Environment and Tourism*, London: Routledge.

Holloway, J. (2002) *The Business of Tourism* (6th edn), Harlow: Pearson Education.

Holt, D. (1995) How consumers consume: a typology of consumption practices, *Journal of Consumer Research* 22 (June): 1–16.

Hoogvelt, A. (1997) *Globalisation and the Postcolonial World*, London: Macmillan.

Horan, J. (2002) Indigenous wealth and development: micro-credit schemes in Tonga, *Asia Pacific Viewpoint* 43(2): 205–221.

Howie, F. (2003) *Managing the Tourist Destination*, London: Thomson.

Høyer, K. (2000) Sustainable tourism or sustainable mobility? The Norwegian case, *Journal of Sustainable Tourism* 8(2): 147–160.

Hughes, G. (1995) Authenticity in tourism, *Annals of Tourism Research* 22(4): 781–803.

Hunter, C. (1995) On the need to re-conceptualise sustainable tourism development, *Journal of Sustainable Tourism* 3(3): 155–165.

Hunter, C. and Green, H. (1995) *Tourism and the Environment: A Sustainable Relationship?*, London: Routledge.

Huybers, T. (ed.) (2007) *Tourism in Developing Countries*, Cheltenham: Edward Elgar.

IFTO (1994) *Planning for Sustainable Tourism: The ECOMOST Project*, Lewes: International Federation of Tour Operators.

Iliau, R. (1997) *The Changing Role of Women in Tonga*, unpublished MA Dissertation, University of Auckland.

IMF (2006) Poverty Reduction Strategy Papers. www.imf.org, accessed 1 August 2006.

India eNews (2006) Bengal to woo foreign investment in tourism, www.indiaenews.com, accessed 17 August 2006.

Inskeep, E. (1991) *Tourism Planning: An Integrated and Sustainable Development Approach*, New York: Van Nostrand Reinhold.

Inskeep, E. and Kallenberger, M. (1992) *An Integrated Approach to Resort Development: Six Case Studies*, Madrid: World Tourism Organisation.

InterContinental (2006) *Group at a Glance Fact Sheet*, http://www.ihgplc.com, accessed 27 September 2006.

Ioannides, D. and Debbage, K. (1998) Neo-Fordism and flexible specialisation, in D. Ioannides and K. Debbage (eds) *The Economic Geography of the Tourist Industry*, London: Routledge, pp. 99–122.

IRN (2002) *Current Developments in the UK Outbound Travel Industry*, Hampton: IRN Research.

Issa, J. and Jayawardena, C. (2003) The 'all-inclusive' concept in the Caribbean, *International Journal of Contemporary Hospitality Management* 15(3): 167–171.

IUCN (1980) *World Conservation Strategy: Living Resources Conservation for Sustainable Development*, Gland, Switzerland: World Conservation Union.

IUCN (1991) *Caring for the Earth: A Strategy for Sustainable Living*, Gland, Switzerland: World Conservation Union.

Jackson, G. and Morpeth, N. (1999) Local Agenda 21 and community participation in tourism policy and planning: future or fallacy, *Current Issues in Tourism* 2(1): 1–38.

Jafari, J. (1989) Sociocultural dimensions of tourism: an English language literature review, in J. Bystrzanowski (ed.) *Tourism as a Factor of Change: A Sociocultural Study*, Vienna: Vienna Centre, pp. 17–60.

Jamal, T. and Getz, D. (1995) Collaboration theory and community tourism planning, *Annals of Tourism Research* 22(1): 186–204.

Jamal, T. and Jamrozy, U. (2006) Collaborative networks and partnerships for integrated destination management, in D. Buhalis and C. Costa (eds) *Tourism Management, Dynamics, Trends, Management and Tools*, London: Elsevier Butterworth Heinemann, pp. 164–172.

Jamal, T., Borgers, M. and Stronza, A. (2006) The institutionalisation of ecotourism: Certification, cultural equity and praxis, *Journal of Ecotourism* 5(3): 145–175.

James, P. (2006) *Globalism, Nationalism and Tribalism, Bringing Theory Back In*, London: Sage.

JCP Inc (1987) *Nusa Tenggara, Tourism Development Plan for Lombok*, Tokyo: JCP Inc. Planners, Architects and Consulting Engineers.

Jenkins, C. (1980) Tourism policies in developing countries: a critique, *International Journal of Tourism Management* 1(1): 22–29.

Jenkins, C. (1991) Development strategies, in L. Lickorish, A. Jefferson, J. Bodlender and C. Jenkins (eds) *Developing Tourist Destinations*, London: Longman, pp. 59–118.

Johansson, Y. and Diamantis, D. (2004) Ecotourism in Thailand and Kenya: a private sector perspective, in D. Diamantis (ed.) *Ecotourism Management and Assessment*, London: Thomson, pp. 298–312.

Jones, S. (2005) Community-based ecotourism: the significance of social capital, *Annals of Tourism Research* 32(2): 303–324.

Kepp, M. (2005) Beach blanket Brazil: global hotel chains see tourism on the rise in Brazil and are spending big bucks now, *Latin Trade*, March, http://findarticles.com/p/articles/mi_m0BEK/is_3_13/ai_n13619929, accessed 28 September 2006.

Kimball, A. (2006) *Risky Trade, Infections Disease in the Era of Global Trade*, Aldershot: Ashgate Publishing.

Knowles, T., Diamantis, D. and El-Mourabi, J. (2001) *The Globalisation of Tourism: A Strategic Perspective*, London: Continuum.

Knox, P., Agnew, J. and McCarthy, L. (2003) T*he Geography of the World Economy* (4th edn), London: Edward Arnold.

Koch, E. and Massyn, P. (2001) South Africa's domestic tourism sector: promises and problems, in K. Ghimire (ed.) *The Native Tourist: Mass Tourism within Developing Countries*, London: Earthscan, pp. 142–171.

Kousis, M. (2000) Tourism and the environment: a social movements perspective, *Annals of Tourism Research* 27(2): 468–489.

Krippendorf, J. (1986) Tourism in the system of industrial society, *Annals of Tourism Research* 13(4): 517–532.

Krippendorf, J. (1987) *The Holiday Makers*, Oxford: Heinemann.

Kusluvan, S. and Karamustafa, K. (2001) Multinational hotel development in developing countries: an exploratory analysis of critical policy issues, *International Journal of Tourism Research* 3: 179–197.

Lai, K., Li, Y. and Feng, X. (2006) Gap between tourism planning and implementation: a case of China, *Tourism Management* 27: 1171–1180.

Lane, B. (1990) Sustaining host areas, holiday makers and operators alike, in *Conference Proceedings, Sustainable Tourism Development Conference*, Queen Margaret College, Edinburgh, November.

Lane, J. (2005) *Globalization and Politics*, Aldershot: Ashgate Publishing.

Lea, J. (1988) *Tourism and Development in the Third World*, London: Routledge.

Leinbach, T. and Bowen, J. (2004) Airspaces: air transport, technology and society, in S. Brunn, S. Cutter and J. Harrington (eds) *Geography and Technology*, London: Kluwer Academic, pp. 285–313.

Leiper, N. (1979) The framework of tourism, *Annals of Tourism Research* 6(1): 390–407.

Leisen, B. (2001) Image segmentation: the case of a tourism destination, *Journal of Services Marketing* 15(1): 49–66.

Lett, J. (1989) Epilogue to touristic studies in anthropological perspective, in V. Smith (ed.) *Hosts and Guests: The Anthropology of Tourism* (2nd edn), Philadelphia: University of Pennsylvania Press, pp. 265–279.

Li, Y. (2004) Exploring community tourism in China: the case of Nanshan Cultural Tourism Zone, *Journal of Sustainable Tourism* 12(3): 175–193.

Lickorish, L. (1991) International agencies, in L. Lickorish, A. Jefferson, J. Bodlender and C. Jenkins (eds) *Developing Tourism Destinations: Policies and Perspectives*, Harlow: Longman, pp. 147–165.

Lipscomb, A. (1998) Village-based tourism in the Solomon Islands: impediments and impacts, in E. Laws, B. Faulkner and G. Moscardo (eds) *Embracing and Managing Change in Tourism*, London: Routledge, pp. 185–201.

Liu, A. and Wall, G. (2006) Planning tourism employment: a developing country perspective, *Tourism Management* 27: 159–170.

Lloyd, K. (2004) Tourism and transitional geographies: mismatched expectations of tourism investment in Vietnam, *Asia Pacific Viewpoint* 45(2): 197–215.

Long, V. (1993) Techniques for socially sustainable tourism development: lessons from Mexico, in J. Nelson, R. Butler and G. Wall (eds) *Tourism and Sustainable Development: Monitoring, Planning, Managing*, Waterloo: Heritage Resources Centre Joint Publication No. 1, University of Waterloo, pp. 201–219.

Ludwig, D., Hilborn, R. and Walters, C. (1993) Uncertainty, resource exploitation, and conservation: lessons from history, *Science* 269(5104):17, 36.

Lury, C. (1996) *Consumer Culture*, Cambridge: Polity Press.

Mabogunje, A. (1980) *The Development Process: A Spatial Perspective*, London: Hutchinson.

MacCannell, D. (1989) The *Tourist: A New Theory of the Leisure Class* (2nd edn), New York: Shocken Books.

McCool, S. and Lime, D. (2001) Tourism carrying capacity: tempting fantasy or useful reality?, *Journal of Sustainable Tourism* 9(5): 372–388.

McCormick, J. (1995) *The Global Environmental Movement*, Chichester: John Wiley & Sons.

McElroy, J. and de Albuquerque, K. (2002) Problems for managing sustainable tourism in small islands, in Y. Apostlolpoulos and D. Gayle (eds) *Island Tourism and Sustainable Development: Caribbean, Pacific and Mediterranean Experiences*, Westport: Praeger, pp. 15–34.

McGehee, N. and Andereck, L. (2004) Factors predicting rural residents' support of tourism, *Journal of Travel Research* 43: 131–140.

MacKenzie, M. (2006) Temples doomed by tourism, *Independent Online*, http://travel.independent.co.uk/news_and_advice/article1090291.ece.

McKercher, B. (1993) Some fundamental truths about tourism: understanding tourism's social and environmental impacts, *Journal of Sustainable Tourism* 1(1): 6–16.

MacLellan, R., Dieke, P. and Thopo, B. (2000) Mountain tourism and public policy in Nepal, in P. Godde, M. Price and F. Zimmerman (eds) *Tourism and Development in Mountain Regions*, Wallingford: CABI, pp. 173–197.

Macleod, D. (2004) *Tourism, Globalisation and Cultural Change: An Island Community Perspective*, Clevedon: Channel View Publications.
McMichael, P. (2004) *Development and Social Change: A Global Perspective* (3rd edn), London: Pine Forge Press.
Macnaught, T. (1982) Mass tourism and the dilemmas of modernization in Pacific island communities, *Annals of Tourism Research* 9(3): 359–381.
Madrigal, R. and Kahle, L. (1994) Predicting vacation activity preferences on the basis of value-system segmentation, *Journal of Travel Research* 32(3): 22–28.
Malta Tourism Authority (2006) Malta Tourism Authority Strategic Plan 2006–2009, http://www.mta.com.mt/index.pl/mta_news, accessed 2 October 2006.
Mann, M. (1986) *The Sources of Social Power, Volume 1: A History of Power from the Beginning to A.D. 1760*, Cambridge: Cambridge University Press.
Mann, M. (2000) *The Community Tourism Guide*, London: Earthscan.
Martell, L. (1994) *Ecology and Society*, Cambridge: Polity Press.
Martin de Holan, P. and Phillips, N. (1997) Sun, sand and hard currency: tourism in Cuba, *Annals of Tourism Research* 24(4): 777–795.
Marwick, M. (2000) Golf tourism development, stakeholders, differing discourses and alternative agendas: the case of Malta, *Tourism Management* 21(5): 515–524.
Mason, P. and Mowforth, M. (1995) *Codes of Conduct in Tourism*, Occasional Papers in Geography No. 1, University of Plymouth: Department of Geographical Sciences.
Mason, P. and Mowforth, M. (1996) Codes of conduct in tourism, *Progress in Tourism and Hospitality Research* 2(2): 151–164.
Mathieson, A. and Wall, G. (1982) *Tourism: Economic, Physical and Social Impacts*, Harlow: Longman.
Mbaiwa, J. (2005) Enclave tourism and its socio-economic impacts in the Okavango Delta, Botswana, *Tourism Management* 26: 157–172.
Mieczkowski, Z. (1995) *Environmental Issues of Tourism and Recreation*, Lanham, MD: University Press of America.
MIGA (2006) MIGA: Supporting Tourism and Hospitality Investments, Tourism and Hospitality Brief, www.miga.org/documents/touris0.6pdf, accessed 22 September.
Milne, S. and Ewing, G. (2004) Community participation in Caribbean tourism: problems and prospects, in D. Duval (ed.) *Tourism in the Caribbean: Trends, Development, Prospects*, London: Routledge, pp. 205–217.
Milne, S. and Gill, K. (1998) Distribution technologies and destination development: myths and realities, in D. Ioannides and K. Debbage (eds) *The Economic Geography of the Tourist Industry*, London: Routledge, pp.123–138.
Milner, H. (1991) The assumption of anarchy in international relations: a critique, *Review of International Studies* 17(1): 67.
Mintel (1994) *The Green Consumer I: The Green Conscience*, London: Mintel International.
Mintel (2007) *Green and Ethical Consumers*, London: Mintel International.
Mishan, E. (1969) *The Costs of Economic Growth*, Harmondsworth: Penguin.
Mitchell, B. (1997) *Resource and Environmental Management*, Harlow: Addison Wesley Longman.
Mitchell, D. (2000) *Cultural Geography a Critical Introduction*, Oxford: Blackwell.
Miossec, J. M. (1976) Elements pour une theorie de l'éspace touristique. *Les Cahiers du Tourisme* C-36. CHET, Aix-en-Provence: Mintel.
Momsen, J. (2004) *Gender and Development*, London: Routledge.
Mosley, P. and Toye, J. (1988) The design of Structural Adjustment Programmes, *Development Policy Review* 6(4): 395–413.

Mowforth, M. and Munt, I. (1998) *Tourism and Sustainability: New Tourism in the Third World*, London: Routledge.

Mowforth, M. and Munt, I. (2003) *Tourism and Sustainability: Development and New Tourism in the Third World* (2nd edn), London: Routledge.

Mowl, G. (2002) Tourism and the environment, in R. Sharpley (ed.) *The Tourism Business: An Introduction*, Sunderland: Business Education Publishers, pp. 219–242.

Murphy, P. (1985) *Tourism: A Community Approach*, New York: Routledge.

Nash, D. (1989) Tourism as a form of imperialism, in V. Smith (ed.) *Hosts and Guests: The Anthropology of Tourism* (2nd edn), Philadelphia: University of Pennsylvania Press, pp. 37–52.

Nepal, S. (2000) Tourism in protected areas: the Nepalese Himalaya, *Annals of Tourism Research* 27(3): 661–681.

Newell, P. (2005) Environment, in P. Burnell and V. Randall (eds) *Politics in the Developing World*, Oxford: Oxford University Press, pp. 221–236.

Newsome, D., Moore, S. and Dowling, R. (2000) *Natural Area Tourism: Ecology, Impacts and Management*, Clevedon: Channel View Publications.

Noronha, L., Siqueira, A., Sreekesh, S., Qureshy, L. and Kazi, S. (2002) Goa: tourism migrations and ecosystem transformations, *Ambio* 31(4): 295–302.

Nozick, M. (1993) Five principles of sustainable community development, in E. Shragge (ed.) *Community Economic Development: In Search of Empowerment and Alteration*, Montreal: Black Rose Books, pp. 18–43.

O'Connor, P., Buhalis, D. and Frew, A. (2001) The transformation of tourism distribution channels through information technology, in D. Buhalis and E. Laws (eds) *Tourism Distribution: Channels Practices, Issues and Transformations*, London: Continuum, pp. 315–331.

OECD (1981) *The Impact of Tourism on the Environment*, Paris: Organisation for Economic Co-operation and Development.

OECD (2001) *Review of the National Tourism Policy in Mexico - Conclusions*, www.oecd.org/dataoecd/44/30/33650504.pdf, accessed 19 September 2006.

O' Grady, R. (1980) *Third World Stopover*, Geneva: World Council of Churches.

Olesen. A. (2006) Train route opens travel to remote Tibet, *Baltimoresun.com*, www.baltimoresun.com, accessed 14 August 2006.

Oneworld (2006) www.oneworld.com, accessed 8 August 2006.

Opperman, M. and Chon, K. (1997) *Tourism in Developing Countries*, London: International Thomson Business Press.

O'Reilly, K. (2003) When is a tourist? The articulation of tourism and migration in Spain's Costa del Sol, *Tourist Studies* 3(3): 301–317.

Ottaway, M. (2005) Civil society, in P. Burnell and V. Randall (eds) *Politics in the Developing World*, Oxford: Oxford University Press. pp. 120–135.

Page, S. and Connell, J. (2006) *Tourism: A Modern Synthesis* (2nd edn), London: Thomson.

Pal, M. (1994) Constraints facing the small-scale informal sectors in developing economies, *Ecodecision*, (autumn): 79–81.

Palma, G. (1995) Underdevelopment and Marxism: from Marx to the theories of imperialism and dependency, in R. Ayers (ed.) *Development Studies: An Introduction Through Selected Readings*, Dartford: Greenwich University Press, pp. 161–210.

Palmer, N. (2006) Economic transition and the struggle for local control in ecotourism development: the case of Kyrgyzstan, *Journal of Ecotourism* 5(1 and 2): 40–61.

Papatheodorou, A. (2006) Liberalisation and deregulation for tourism: implications for competition, in C. Costa and D. Buhalis (eds) *Tourism Management Dynamics, Trends, Management and Tools*, London: Elsevier Butterworth Heinemann, pp. 68–77.

Parinello, G. (1993) Motivation and anticipation in post-industrial tourism, *Annals of Tourism Research* 20(2): 233–249.

Pattulo, P. with Minelli, O. (2006) *Ethical Travel Guide*, London: Earthscan.

Pearce, D. (1989) *Tourist Development* (2nd edn), Harlow: Longman.

Pearce, D., Markandya, A. and Barbier, E. (1989) *Blueprint for a Green Economy*, London: Earthscan.

Pearce, P. (1992) Fundamentals of tourist motivation, in D. Pearce and R. Butler (eds) *Tourism Research: Critiques and Challenges*, London: Routledge, pp. 113–134.

Pearce, P. (2005) *Tourist Behaviour Themes and Conceptual Schemes*, Clevedon: Channel View Publications.

Peet, R. with Hartwick, E. (1999) *Theories of Development*, London: Guilford Press.

Perkins, J. (2004) *Confessions of an Economic Hit Man*, London: Penguin Books.

Peterson, R. (1979) Revitalizing the culture concept, *Ann. Rev. Sociol.* 5: 137–166.

Pigram, J. (1990) Sustainable tourism – policy considerations, *Journal of Tourism Studies* 1(2): 2–9.

Pike, S. (2002) Destination image analysis – a review of 142 papers from 1973 to 2000, *Tourism Management* 23(4): 541–549.

Plog, S. (1977) Why destinations rise and fall in popularity, in E. Kelly (ed.) *Domestic and International Tourism*, Wellesley, MA.: Institute of Certified Travel Agents.

Plummer, R. and Fitzgibbon, J. (2004) Some observations on the terminology in co-operative environmental management, *Journal of Environmental Management* 70(1): 63–72.

Poon, A. (1989) Competitive strategies for a 'new tourism', in C. Cooper (ed.) *Progress in Tourism, Recreation and Hospitality Management, Vol 1*, London: Belhaven Press.

Poon, A. (1993) *Tourism, Technology and Competitive Strategies*, Wallingford: CAB International.

Potter, D. (2000) Democratisation, 'good governance' and development in T. Allen and A. Thomas (eds) *Poverty and Development into the 21st Century*, Oxford: Oxford University Press, pp. 365–382.

Potter, R. B. (1995) Urbanisation and development in the Caribbean, *Geography* 80: 334–341.

Potter, R. B. (2002) Theories, strategies and ideologies of development, in V. Desai and R. B. Potter (eds) *The Companion to Development Studies*, New York: Oxford University Press, pp. 61–65.

Potter, R., Binns, T., Elliot, J. and Smith, D. (1999) *Geographies of Development*, Harlow: Prentice Hall.

PPT (2004) Pro-poor tourism info-sheets, sheet No. 9: Tourism in Poverty Reduction Strategy Papers (PRSPs), www.propoortourism.org.uk, accessed 8 August 2006.

Prasad, E., Rogoff, K., Wei, S. and Kose, A. (2003) *Effects of Financial Globalisation on Developing Countries, Some Empirical Evidence*, International Monetary Fund Occasional Paper 220, 9 September, International Monetary Fund, www.imf.org/external/pubs/nft/op/220/index.htm, accessed 8 May 2007.

Preston, P. (1996) *Development Theory: An Introduction*, Oxford: Blackwell.

Preston-Whyte, R. and Watson, H. (2005) Nature tourism and climate change in Southern Africa, in C. M. Hall and J. Higham, (eds) *Tourism, Recreation and Climate Change*, Clevedon: Channel View Publications, pp. 130–142.

Proops, J. and Wilkinson, D. (2000) Sustainability, knowledge, ethics and the law, in M. Redclift (ed.) *Sustainability, Life Chances and Livelihoods*, London: Routledge, pp. 17–34.

Rahnema, M. and Bawtree, V. (eds) (1997) *The Post-development Reader*, London: Zed Books.

Rain, D. and Brooker-Gross, S. (2004) A world on demand: geography of the 24-hour global

TV news, in S. Brunn, S. Cutter and J. Harrington (eds) *Geography and Technology*, London: Kluwer Academic, pp. 315–337.

Rakodi, C. (1995) Poverty lines or household strategies?, *Habitat International* 19(4): 407–426.

Randall, V. (2005) Analytical approaches to the study of politics in the developing world, in P. Burnell and V. Randall (eds) *Politics in the Developing World*, Oxford: Oxford University Press, pp. 9–23.

Rapley, J. (2002) *Understanding Development Theory and Practice in the Third World*, London: Lynne Reinner.

Ravallion, M. (2004) *Pro-poor Growth: A Primer*, Washington, DC: World Bank.

Ray, C. (1998) Culture, intellectual property and territorial rural development, *Sociologia Ruralis* 38: 3–20.

Redclift, M. (1987) *Sustainable Development: Exploring the Contradictions*, London: Routledge.

Redclift, M. (2000) Introduction, in M. Redclift (ed.) *Sustainability, Life Chances and Livelihoods*, London: Routledge, pp. 1–13.

Reid, D. (1995) *Sustainable Development: An Introductory Guide*, London: Earthscan.

Reid, D. (2003) *Tourism, Globalization and Development: Responsible Tourism Planning*, London: Pluto Press.

Reisinger, Y. and Turner, L. (2003) *Cross-cultural Behaviour in Tourism Concepts and Analysis*, Oxford: Butterworth-Heinemann.

Richards, G. and Hall, D. (2000a) The community: a sustainable concept in tourism development?, in D. Hall and G. Richards (eds) *Tourism and Sustainable Community Development*, London: Routledge, pp. 1–13.

Richards, G. and Hall, D. (2000b) Conclusions, in D. Hall and G. Richards (eds) *Tourism and Sustainable Community Development*, London: Routledge, pp. 297–306.

Robbins, P. (2001) *Greening the Corporation: Management Strategies and the Environmental Challenge*, London: Earthscan.

Roberts, S. (2002) Global regulation and trans-state organisation, in R. Johnston, P. Taylor and M. Watts (eds) *Geographies of Global Change: Remapping the World*, Oxford: Blackwell, pp. 143–157.

Robins, K. (1997) What in the world is going on?, in P. du Gay (ed.) *Production of Culture/Cultures of Production*, London: Sage, pp. 11–66.

Rodrik, D. (2005) Feasible globalisations, in M. Weinstein (ed.) *Globalization: What's New?*, New York: Columbia University Press, pp. 196–213.

Rogerson, C. (2006) Pro-poor local economic development in South Africa: the role of pro-poor tourism, *Local Environment* 11(1): 37–60.

Rokeach (1973) *The Open and Closed Mind*, New York: Basic Books.

Rostow, W. (1967) *The Stages of Economic Growth: A Non-Communist Manifesto* (2nd edn), Cambridge: Cambridge University Press.

Rothman, J., Erlich, M. and Tropman, J. E. (1995) *Strategies of Community Intervention* (5th edn), Itasca, IL: F. E. Peacock.

Routledge, P. (2001) 'Selling the rain', resisting the sale: resistant identities and the conflict over tourism in Goa, *Social and Cultural Geography* 2(2): 221–240.

Ryan, C. (1991) *Recreational Tourism: A Social Science Perspective*, London: Routledge.

Ryan, C. (1997) The chase of a dream, the end of a play, in C. Ryan (ed.) *Tourist Experience: A New Introduction*, London: Cassell, pp. 1–24.

Ryan, C. (2005) Introduction: tourist–host nexus – research considerations, in C. Ryan and M. Aicken (eds) *Indigenous Tourism; The Commodification and Management of Culture*, London: Elsevier, pp. 1–11.

Ryan, C. and Hall, C. M. (2001) *Sex Tourism: Marginal People and Liminalities*, London: Routledge.

Sachs, J. (2005) *The End of Poverty: Economic Possibilities for Our Time*, New York: Penguin Books.

Sachs, W. (1996) Introduction, in W. Sachs (ed.) *The Development Dictionary: A Guide to Knowledge and Power*, London: Zed Books, pp. 2–5.

Said, E. (1978) *Orientalism*, London: Routledge.

Salem, N. (1994) Water rights, *Tourism in Focus* (Tourism Concern) 17: 4–5.

Saul, J. R. (2005) *The Collapse of Globalism and the Reinvention of the World*, Toronto: Viking Canada.

Schemo, J. (1995) Galapagos Island Journal: Homo sapiens at war on Darwin's peaceful island, *New York Times*, 15 August, retrieved from www.nytimes.com.

Scheyvens, R. (2002) *Tourism for Development: Empowering Communities*. London: Prentice Hall.

Scheyvens, R. (2003) Local involvement in managing tourism, in S. Singh, D. Timothy and R. Dowling (eds) *Tourism in Destination Communities*, Wallingford: CABI, pp. 229–252.

Schilcher, D. (2007) Growth versus equity: the continuum of pro-poor tourism and neoliberal governance, *Current Issues in Tourism* 10(2 and 3): 166–193.

Schmidt, H. (1989) What makes development?, *Development and Cooperation* 6:19–26.

Scholte, J. (2005) *Globalisation: A Critical Introduction*, New York: Palgrave.

Schumacher, E. (1974) *Small is Beautiful: A Study of Economics as if People Mattered*, London: Abacus.

Schuurman, F. (1996) Introduction: development theory in the 1990s, in F. Schuurman (ed.) *Beyond the Impasse: New Direction in Development Theory*, London: Zed Books, pp. 1–48.

Seckelman, A. (2002) Domestic tourism – a chance for regional development in Turkey?, *Tourism Management* 23(1): 85–92.

Seers, D. (1969) The meaning of development, *International Development Review* 11(4): 2–6.

Sen, A. (1999) *Development as Freedom*, New York: Anchor Books.

Shackley, M. (1996) *Wildlife Tourism*, London: International Thomson Business Press.

Shah, K. (2000) *Tourism, the Poor and Other Stakeholders: Asian Experience, ODI Fair-Trade Tourism Paper*, London: ODI.

Sharpley, R. (1994) *Tourism, Tourists and Society*, Huntingdon, Cambridgeshire: Elm Publications.

Sharpley, R. (2000) Tourism and sustainable development: exploring the theoretical divide, *Journal of Sustainable Tourism* 8(1): 1–19.

Sharpley, R. (2001) Tourism in Cyprus: challenges and opportunities, *Tourism Geographies* 3(1): 64–86.

Sharpley, R. (2002) *The Tourism Business: An Introduction*, Sunderland: Business Education Publishers.

Sharpley, R. (2003) *Tourism, Tourists and Society* (3rd edn), Huntingdon: Elm Publications.

Sharpley, R. (2005) The tsunami and tourism: a comment, *Current Issues in Tourism* 8(4), 344–349.

Sharpley, R. (2006a) Tourism in The Gambia: 10 years on, Paper presented at the Cutting Edge in Tourism Research Conference, University of Surrey, June.

Sharpley, R. (2006b) Ecotourism: a consumption perspective, *Journal of Ecotourism* 5(1 and 2): 7–22.

Sharpley, R. and Telfer, D. J. (eds) (2002) *Tourism and Development Concepts and Issues*, Clevedon: Channel View Publications.

Sharpley, R., Sharpley, J. and Adams, J. (1996) Travel advice or trade embargo: the impacts and implications of official travel advice, *Tourism Management* 17(1): 1–7.

Shaw, B. and Shaw, G. (1999) 'Sun, sand and sales': enclave tourism and local entrepreneurship in Indonesia, *Current Issues in Tourism* 2(1): 68–81.

Silver, I. (1993) Marketing authenticity in Third World countries, *Annals of Tourism Research* 20(2): 302–318.

Simpson, B. (1993) Tourism and tradition: from healing to heritage, *Annals of Tourism Research* 20(2): 164–181.

Simpson, K. (2004) 'Doing development': the gap year volunteer-tourists and a popular practice of development, *Journal of International Development* 16: 681–692.

Singh, S. (2001) Indian tourism: policy, performance and pitfalls, in D. Harrison (ed.) *Tourism and the Less Developed World: Issues and Case Studies*, Wallingford: CABI, pp. 137–149.

Singh, S., Timothy, D. and Dowling, R. (2003) Tourism and destination communities, in S. Singh, D. Timothy and R. Dowling (eds) *Tourism in Destination Communities*, Wallingford: CABI, pp. 3–18.

Sklair, L. (1995) *Sociology of the Global System*. Baltimore, MD: The Johns Hopkins University Press.

Sky Team (2006) Sky Team, www.skyteam.com/skyteam, accessed 8 August 2006.

Smith, M. (2007) Cultural tourism in a changing world, *Tourism: The Journal for the Tourism Industry* 1(1): 18–19.

Smith, S. (1994) The tourism product, *Annals of Tourism Research* 21(3): 582–595.

Smith, V. (ed.) (1977) *Hosts and Guests: The Anthropology of Tourism* (1st edn), Philadelphia: University of Pennsylvania Press.

Smith, V. (ed.) (1989) *Hosts and Guests: The Anthropology of Tourism* (2nd edn), Philadelphia: University of Pennsylvania Press.

Smith, V. (1996) Indigenous tourism; the four Hs, in R. Butler and T. Hinch (eds) *Tourism and Indigenous Peoples*, London: International Thomson Business Press, pp. 283–307.

Smith, V. and Eadington, W. (eds) (1992) *Tourism Alternatives: Potentials and Problems in the Development of Tourism*, Philadelphia: University of Pennsylvania Press.

Sofreavia in association with PT. Asana Wirasta Setia and PT. Desigras (1993) *Feasibility Study for Airport Development in Lombok, Master Plan Executive Summary*.

Solomon, M. (1994) *Consumer Behaviour: Buying, Having, Being* (2nd edn), Needham Heights, MA: Allyn & Bacon

Southgate, C. (2006) Ecotourism in Kenya: the vulnerability of communities, *Journal of Ecotourism* 5(1 and 2): 80–96.

Southgate, C. and Sharpley, R. (2002) Tourism, development and the environment, in R. Sharpley and D. Telfer (eds) *Tourism and Development: Concepts and Issues*, Clevedon: Channel View publication, pp. 231–262.

Spenceley, A. (2004) Responsible nature-based tourism planning in South Africa and the commercialisation of Kruger National Park, in D. Diamantis (ed.) *Ecotourism Management and Assessment*, London: Thomson, pp. 267–280.

Star Alliance (2006) Star Alliance facts and figures, www.staralliance.com, accessed 8 August 2006.

Starmer-Smith, C. (2004) Eco-friendly tourism on the rise. *Daily Telegraph Travel*, 6 November, p. 4.

Steer, A. and Wade-Gery, W. (1993) Sustainable development: theory and practice for a sustainable future, *Sustainable Development* 1(3): 23–35.

Stoddart, H. and Rogerson, C. (2004) Volunteer tourism: the case of Habitat for Humanity South Africa, *GeoJournal* 60: 311–318.

Streeten, P. (1977) The basic features of a basic needs approach to development, *International Development Review* 3: 8–16.

Stubbs, R. and Underhill, G. (eds) (2006) *Political Economy and the Changing Global Order* (3rd edn), Don Mills, Ontario: Oxford University Press.

Taylor, P., Watts, M. and Johnston, R. (2002) Geography/globalisation, in R. Johnston, P. Taylor and M. Watts (eds) *Geographies of Global Change: Remapping the World*, Oxford: Blackwell, pp. 1–17.

Tearfund (2000) *Tourism – An Ethical Issue. Market Research Report*, Teddington: Tearfund.

Telfer, D. J. (1996) Food purchases in a five-star hotel: a case study of the Aquila Prambanan Hotel, Yogyakarta, Indonesia, *Tourism Economics* 2(4): 321–338.

Telfer, D. J. (2000) Agritourism – a path to community development? The case of Bangunkerto, Indonesia, in D. Hall and G. Richards (eds) *Tourism and Sustainable Community Development*, London: Routledge, pp. 242–257.

Telfer, D. J. (2001) Tourism and Community Development in a Biosphere: Sierra Del Rosario, Cuba, unpublished paper.

Telfer, D. J. (2002a) The evolution of tourism and development theory, in R. Sharpley and D. J. Telfer (eds) *Tourism and Development: Concepts and Issues*, Clevedon: Channel View Publications, pp. 35–78.

Telfer, D. J. (2002b) Tourism and regional development issues, in R. Sharpley and D. J. Telfer (eds) *Tourism and Development: Concepts and Issues*, Clevedon: Channel View Publications, pp. 112–148.

Telfer, D. J. (in press) Development studies and tourism, in M. Robinson and T. Jamal (eds) *Handbook of Tourism Studies*, London: Sage.

Telfer, D. J. and Wall, G. (1996) Linkages between tourism and food production, *Annals of Tourism Research* 23(3): 635–653.

Telfer, D. J. and Wall, G. (2000) Strengthening backward economic linkages: local food purchasing by three Indonesian hotels, *Tourism Geographies* 2(4): 421–447.

Teo, P. (2002) Striking a balance for sustainable tourism: implications of the discourse on globalisation, *Journal of Sustainable Tourism* 10(6): 459–474.

The Economist (2007) Globalisation's offspring, *The Economist* 383(8523): 11.

Thomas, A. (2000) Poverty and the 'end of development', in T. Allen and A. Thomas (eds) *Poverty and Development into the 21st Century*, Oxford: Oxford University Press, pp. 3–22.

Timothy, D. (1998) Cooperative tourism planning in a developing destination, *Journal of Sustainable Tourism* 6(1): 52–68.

Timothy, D. (1999) Participatory planning: a view of tourism in Indonesia, *Annals of Tourism Research* 26(2): 371–391.

Timothy, D. (2000) Tourism planning in Southeast Asia: bringing down borders through cooperation, in K. Chon (ed.) *Tourism in Southeast Asia: A New Direction*. New York: The Haworth Hospitality Press, pp. 21–38.

Timothy, D. (2004) Tourism and supranatioanlism in the Caribbean, in D. Duval (ed.) *Tourism in the Caribbean Trends, Development, Prospects*, London: Routledge, pp. 119–135.

Timothy, D. (2005) *Shopping Tourism, Retailing and Leisure*, Clevedon: Channel View Publications.

Timothy, D. and Boyd, S. (2003) *Heritage Tourism*, London: Prentice Hall.

Todaro, M. (1997) *Economic Development* (6th edn), Harlow: Addison-Wesley.

Todaro, M. (2000) *Economic Development* (7th edn), Harlow: Addison-Wesley.

Torres, R. (2002) Cancún's tourism development from a Fordist spectrum of analysis, *Tourist Studies* 2(1): 87–116.

Torres, R. (2003) Linkages between tourism and agriculture in Mexico, *Annals of Tourism Research* 30(3): 546–566.

Torres, R. and Momsen, J. H., (2004) Challenges and potential of linking tourism and

agriculture to achieve pro-poor tourism objectives, *Progress in Development Studies* 4(4): 294–318.
Tosun, C. (1999) An analysis of the economic contribution of inbound international tourism in Turkey, *Tourism Economics* 5(3): 217–250.
Tosun, C. (2000) Limits to community participation in the tourism development process in developing countries, *Tourism Management* 21: 613–633.
Tourism Concern (2006) Global hotel chain's claims of responsible development are a 'greenwash' says pressure group, Tourism Concern Press Release, 25 July, from www.tourismconcern.org.uk/media/2006/Hilton%20and%20CSR%20July%2006.htm, accessed November 2006.
Tourism Concern (n.d.) *Behind the Smile The Tsunami of Tourism*, London: Tourism Concern.
Tourism Thailand (2006) *About TAT*, http://www.tourismthailand.org/about/abouttat.aspx, accessed 14 September.
Transat (2006) Profile, www.airtransit.ca, accessed 11 August 2006.
Tribe, J., Font, X., Griffiths, N., Vickery, R. and Yale, K. (2000) *Environmental Management for Rural Tourism and Recreation*, London: Cassell.
Turner, L. and Ash, J. (1975) *The Golden Hordes: International Tourism and the Pleasure Periphery*, London: Constable.
TVNZ (2006) Coup costing Fiji tourism millions, http://tvnz.co.nz/view/page/411419/930768, accessed 15 February 2007.
United Nations (UN) (1955) *Social Progress Through Community Development*, New York: United Nations.
UN (2003) *Poverty Alleviation Through Sustainable Tourism Development*, New York: United Nations.
UN (2007) Universal Declaration of Human Rights, United Nations, retrieved from www.unhchr.ch/udhr/ 4 July 2007.
UNCTAD (2001) Tourism and development in the Least Developed Countries, *Third UN Conference on the Least Developed Countries*, Las Palmas, CI.
Underhill, G. (2006) Conceptualising the changing global order, in R. Stubbs and G. Underhill (eds) *Political Economy and the Changing Global Order Third Edition*, Don Mills, Ontario: Oxford University Press, pp. 3–23.
UNDP (1992) *Tourism Sector Programming and Policy Development, Output 1, National Tourism Strategy*, New York: United Nations Development Programme.
UNDP (2004) *Human Development Report 2004*, New York: United Nations Development Programme.
UNEP (2002) *Tourism's Three Main Impact Areas*, www.uneptie.org/pc/tourism/sust-tourism/env-3main.htm.
UNEP (2005) *Integrating Sustainability into Business: A Management Guide for Responsible Tour Operations*, Paris: United Nations Environment Programme.
UNEP/WTO (2005) *Making Tourism More Sustainable: A Guide for Policy Makers*, Paris/Madrid: United Nations Environment Programme/World Tourism Organization.
UN-Habitat (2003) *The Challenges of Slums: Global Report on Human Settlements 2003*, London: Earthscan.
UNWTO (2006) *Technical Cooperation, An Effective Tool for Development Assistance*, http://www.world-tourism.org/techcoop/eng/objectives.htm, accessed 24 August.
Uriely, N. (1997) Theories of modern and postmodern tourism, *Annals of Tourism Research* 24(4): 982–985.
Urry, J. (1990) *The Tourist Gaze*, London: Sage.
Urry, J. (1995) *Consuming Places*, London: Routledge.
Urry, J. (2001) *The Tourist Gaze* (2nd edn), London: Sage.

Vargas, C. (2000) Community development and micro-enterprises: fostering sustainable development, *Sustainable Development*, 8: 11–26.

Visit Mexico Press (2006a) Private Investment in Mexico's Tourism Sector Booming, Sectur expects year-end total to surpass US$12 billion, www.visitmexicopress.com?press_release02.ap?pressID=179, accessed 21 September 2006.

Visit Mexico Press (2006b) Fonatur outlines accomplishments and future development strategy at Tiangus, www.visitmexicopress.com?press_release02.ap?pressID=162, accessed 21 September 2006.

Wade, R. (2004) *Governing the Market*, (2nd edn), Princeton, NJ: Princeton University Press.

Wahab, S. and Cooper, C. (eds) (2001) *Tourism in the Age of Globalization*, London: Routledge.

Wall, G. (1993) Towards a tourism typology, in J. G. Nelson, R. W. Butler and G. Wall (eds) *Tourism and Sustainable Development: Monitoring, Planning, Managing*, University of Waterloo; Heritage Resources Centre Joint Publication, pp. 45–58.

Wall, G. and Mathieson, A. (2006) *Tourism Change, Impacts and Opportunities*, Toronto: Pearson Prentice Hall.

Wall, G. and Xie, P. (2005) Authenticating ethnic tourism: Li Dancers' perspectives, *Asia Pacific Journal of Tourism Research* 10(1): 1–21.

Wang, Y. and Wall, G. (2005) Resorts and residents: stress and conservatism in a displaced community, *Tourism Analysis* 10(1): 37–53.

WCED (1987) *Our Common Future*, Oxford: Oxford University Press.

Wearing, S. (2001) *Volunteer Tourism: Experiences that Make a Difference*, New York: CABI.

Weaver, D. (2004) Manifestations of ecotourism in the Caribbean, in D. Duval (ed.) *Tourism in the Caribbean: Trends, Development, Prospects*, London: Routledge, pp. 172–186.

Weinstein, M. (2005) Introduction, in M. Weinstein (ed.) *Globalization What's New?*, New York: Columbia University Press, pp 1–18.

Wheeller, B. (1992) Eco or ego tourism: new wave tourism, *Insights*, Vol III, London: English Tourist Board, D41–44.

Wiarda, H. J. (1988) Toward a nonethnocentric theory of development: alternative conceptions from the Third World. *Journal of Developing Areas* 17, reprinted in C. K. Wilber (ed.) *The Political Economy of Development and Underdevelopment* (4th edn), Toronto: McGraw-Hill, pp. 59–82.

Wilbanks, T. (2004) Geography and technology, in S. Brunn, S. Cutter and J. Harrington (eds) *Geography and Technology*, London: Kluwer Academic, pp.1–16.

Williams, A. and Hall, C. M. (2000) Tourism and migration: new relationships between production and consumption, *Tourism Geographies* 2(1): 5–27.

Wood, K. and House, S. (1991) *The Good Tourist: A Worldwide Guide for the Green Traveller*, London: Mandarin.

Wood, R. (1997) Tourism and the state: ethnic options and constructions of otherness, in M. Picard and R. Woods, *Tourism, Ethnicity and the State in Asian and Pacific Countries*, Honolulu: University of Hawaii Press, pp.1–34.

Wood, R. (2004) Global currents: cruise ships in the Caribbean, in D. Duval (ed.) *Tourism in the Caribbean: Trends, Development, Prospects*, London: Routledge, pp. 152–171.

World Bank (2005) Classification of economies, www.worldbank.org/data/aboutdata/errata03/ Class.htm, accessed April 2005.

World Trade Organisation (2001) Dhoa WTO Ministerial Declaration. www.wto.org, retrieved 31 July 2006.

Wright, R. (2008) *Environmental Science: Toward a Sustainable Future (10th edn), Instructors' Edition*, Upper Saddle River, NJ: Pearson Prentice Hall.

WTO (1980) *Manila Declaration on World Tourism*, Madrid: World Tourism Organization.

WTO (1981) *The Social and Cultural Dimension of Tourism*, Madrid: World Tourism Organization.

WTO (1986) *Village Tourism Development Programme for Nusa Tenggara*, Madrid: World Tourism Organization, UNDP.

WTO (1993) *Sustainable Tourism Development: A Guide for Local Planners*, Madrid: World Tourism Organization.

WTO (1996) *Agenda 21 for the Travel and Tourism Industry: Towards Environmentally Sustainable Development*, Madrid: World Tourism Organization.

WTO (1998) *Tourism – 2020 Vision: Influences, Directional Flows and Key Influences*, Madrid: World Tourism Organization.

WTO (1999) *Sustainable Tourism Development: An Annotated Bibliography*, Madrid: World Tourism Organization.

WTO (2000) *Tourism Highlights 2000* (2nd edn), Madrid: World Tourism Organization.

WTO (2002) *Tourism and Poverty Alleviation*, Madrid: World Tourism Organization.

WTO (2004a) *Tourism Highlights Edition 2004*, Madrid: World Tourism Organization.

WTO (2004b) *Indicators of Sustainable Development for Tourism Destinations: A Guidebook*, Madrid: World Tourism Organization.

WTO (2005a) News Release: International tourism obtains its best results in 20 years, World Tourism Organization, www.world-tourism.org/newsroom/Releases/2005/january/2004numbers.htm, accessed April 2005.

WTO (2005b) *A Historical Perspective of World Tourism*, World Tourism Organization, www.World-tourism.org/facts/trends/historical.htm, accessed 25 May 2006.

WTO (2005c) *World Tourism Barometer 3(2)*, Madrid: World Tourism Organization.

WTO (2006a) *UNWTO World Tourism Barometer* 4(1), World Tourism Organization, www.world-tourism.org/facts/menu.htm, accessed 18 October 2006.

WTO (2006b) Asian outbound tourism takes off, World Tourism News Release, 14 June, retrieved 28 October 2006 from www.world-tourism.org/newsroom/Release/2006/june/asianoutbound.html.

WTO/UNSTAT (1994) *Recommendations on Tourism Statistics*, Madrid: World Tourism Organization.

WTO/WTTC (1996) *Agenda 21 for the Travel and Tourism Industry: Towards Environmentally Sustainable Development*, Madrid: World Tourism Organization/World Travel and Tourism Council.

WTTC (2003) *Blueprint for a New Tourism*, London: World Travel and Tourism Council.

WTTC (2004) *Country League Tables*, London: World Travel and Tourism Council

WTTC (2006b) *Country League Tables*, London: World Travel and Tourism Council, www.wttc.org/frameset2.htm.

Yamamura, T. (2005) Donga Art in Lijinag, China: indigenous culture, local community and tourism, in C. Ryan and M. Aicken (eds) *Indigenous Tourism; The Commodification and Management of Culture*, London: Elsevier, pp. 181–199.

Yiannakis, A. and Gibson, H. (1992) Roles tourists play, *Annals of Tourism Research* 19(3): 287–303.

Young, G. (1973) *Tourism: Blessing or Blight?*, Harmondsworth: Penguin.

Zhang, H., Chong, K. and Ap, J. (1999) An analysis of tourism development policy in modern China, *Tourism Management* 20(4): 471–485.

Zhang, H., Pine, R. and Lam, T. (2005) *Tourism and Hotel Development in China*, New York: The Haworth Hospitality Press.

Index

Page numbers in *italics* refer to tables, plates and boxes.

acceptable change, limits of (LAC) 203
accommodation 67–9
Accor Group 67, 101
accreditation schemes 51
acculturation 75, 126, 197
ACS (Association of Caribbean States) 91
actors in development process 86–92
adaptancy stage, Jafar's Four Platforms of Tourism 27
adventure tourism 25, 74, 147, 152, 157–8, 160, 162, 221; 'hard' and 'soft' 159
advocacy stage, Jafar's Four Platforms of Tourism 27
Aeroflot 61
Agenda 21 30, 122; Local 122–3, 128, 134
Agenda 21 for the Travel and Tourism Industry 123
agriculture 9, 183, 206, 220; Bhutan *37*; and gender 142; impact of tourism 184, 187, 190; linkages 102, 111
air pollution 188
Airbus 59
airlines 65–6, 68–9, 76, 218; air pollution 188; low cost 58, 59, 61, 161
airports 66, 94–5, 102, 112, 175, 187; secondary 59
Airtours (MyTravel) 69
Alexandra Township *139*
Alitalia 61
all-incusive resorts 26, 58, 102, 109, 152; Cancún, Mexico 26, 60, 61, *65*, *87*, 109, 184, 217–18; Cuba 160; development 99–102; limits on 167, 225–6; Sandals, Jamaica *153*
alternative development *12*
alternative tourism 14, 27, 39, 40; criticisms of 41; integrative development and 102–3; vs mass tourism *39–40*; *see also* ecotourism; sustainable tourism
Amir Palace Hotel, Tunisia *101*
Angkor Wat, Cambodia 192
anti-competitive practices 65
APEC (Asian Pacific Economic Cooperation) 124–5
appropriate tourism 26, 28, 148, 164
Arab States *see* Middle East
Argentina: Estancia Santa Susana *118*, *194*; Iguazu Falls *186*
arrivals/receipts 1–2, 5, 8, 17–19, 21, 24
ASEAN (Association of Southeast Asian Nations) 63, 93–4
Asian Development Bank 124
assimilation 74, 75
attitudes towards tourism 120–1
attractions 19, 25, 77, 81, 86; typology 98
Australia 4–5, 141, 152; Ayers Rock (Uluru) 176; National Ecotourism Strategy 50–1
authenticity 108, 126–7, 154
aviation industry *see* airlines
Ayers Rock (Uluru) 176

backpackers 92, 132, 156
backward linkages 19, 111, 183
Bahamas 59, 91; Nassau *68*, *161*

Bahrain 24
Baja California *87*
Baker, S. 122–3, 128, 134, 142–3
balance of payments 17, 110, 180–1, 184, 225
Balearic Islands 53, 97, 101
Bali 3, 94–5, 102, *184*, *197*, 217–18, *227*
Banco Brasil 99–100
Bangkok, Thailand *107*
Bangladesh 61, 63, 138
Bangunkerto village, Indonesia *125*
Barbados 216
battlefield tourism 160
beach resorts 58, 99, *100*, *101*
behaviours: cross-cultural 74–5; *see also* tourist behaviour
Belize 85, 106; ecotourism in 163, 164, *165*
best practice 50
BEST scheme 53
Bhutan 26, *37–8*, 167, 176
Bintan Island, Indonesia 63
Blueprint for New Tourism (WTTC) 50
Botswana 102
bottom-up planning 11, 14
Brazil 88, 99–101, 122, 170, 171
Bretton Woods 62, 71
British Airways 47–8, 49, 50
British Virgin Islands 24
Britton, S. 5, 13, 25, 71–2, 72, 102
Brundtland Report 32, 34–5
Buenos Aires *118*, *194*
bureaucracy 82, 92, 96, 133–4, 218–19
Burma (Myanmar) 1–2, 24, 176
business culture 77–8
Bwindi Impenetrable National Park 135

Cambodia 1–2, 24, 61; Angkor Wat 192
CAMPFIRE programme (Communal Areas Management Programmes for Indigenous Resources) 49
Canada 4–5, 59, 69, 89–90, 220; Global 90 Conference 42
Cancún, Mexico 26, 60, 61, *65*, *87*, 109, 184, 217–18
capitalism 35–6, 93; *see also* globalization
CARE (Cooperation for Assistance and Relief Everywhere) 135
Caribbean and islands 64, 69, 89, 123–4, 220; cruise ships *67*, *68*, 188–9; regional–level organizations 91; Sandals, Jamaica *153*; tourism development dilemma framework 216, 219–20; volunteer tourism 141
carrying capacity 201–2
casinos 67, 91, 92
cautionary stage, Jafar's Four Platforms of Tourism 27
CCC (Caribbean Conservation Corporation) 141
Central African Republic 63
child mortality 9
China 4–5, 9, 15–16, 24, 130; displacement 132; domestic tourism 170, 221–2; ethnic minorities 108–9, 128; Great Wall *21*; SARS outbreak 76; Shanghai 111; Western Region Development Strategy 15–16
Club Mediterranée *153*
Coca-Cola trails 189
codes of conduct 52, 166–7; tour operators' role 169
Coke Line 74
colonialism 26, 71, 142
commodification 76–7, 106, 108–10; of dance masks, Sri Lanka 196
community development and gender 142–3
community-based tourism 124–8; role of NGOs 133–5
community/communities: displacement of 72, 130, 131–2; and environmental linkages 98; interactions with tourism 118–20; involvement/participation 121–4, 128–30; nature of 116–17; power issues 130–3; responses 120–1, 198–200; sustainable development and 121–4
comparative advantage 85, 217
conceptual definition of tourism 5
conflict 15, 58, 75, 82, 86; community responses 117, 122, 128, 144; international 21; of values 156
conservation 32–4
conspicuous consumption 156
consumer culture 156–7
consumption *3*, 221–2; green 163–4; and values 155–6; *see also* tourism demand
control: and scale of development 98; *see also* power
coping mechanisms 207
coral reefs *67*, *165*, 187, 189
corporate social responsibility (CSR) 31, 52–3, 84, 91, 220
Costa Rica 50, 89–90, 105, 141
costs: economic 184–5; opportunity 185; production 63–4, 131, 210, 218, 230
Cozumel, Mexico 109
crime 85, 195, 198
cross-cultural contact 74–6
CRS (corporate social responsibility) 31, 52–3, 84, 91, 220
cruise ships *67*, *68*, 188–9
CSE (Centre for Ecological Support) 134, 135
Cuba 80, 130, 158, 160, 180–1, 196; community-based ecotourism *126*, *127*; Havana *160*; US

embargo on travel to 176; Varadero *18*, *100*, 101–2
Cukier, J. 133, 183, 207
cultural brokers 74
cultural drift 75
cultural globalization 74–8
cultural homogenization 75, 192
cultural impacts 196–7
cultural imperialism 74; *see also* Western imperialism
cultural tourism 76–7, 107–10
culture: business 77–8; commercialization of *see* commodification; consumer 156–7; tourist 117, 215
Cyprus 26, 160, 198, 200

Damaraland Camp, Namibia 40–1
debt 9–10, 71, 78, 99, 110, 111–12, 228
degraded environment 28, 33–4, 142, 171, 189, 198, 211
democracy 14–15, 106, 122, 143–4, 218
democratization 10
demonstration effect 195
dependency 184–5, 220
dependency theory *12*, 13
deregulation: airlines 65–6; financial 59, 61, 64
destination(s) *22*, 24; exotic 1–2, 25, 73, 152, 156, 159–61, 198; factors and characteristics 179; form, function and response 223–4; marketing 167–8; perspective 165–9; planning 97–110, 167
developers 32, 91, 96–7, 112, 113, 120, 131; tourism development dilemma 215, 216–17, 218, 224, 230
developing countries, terminology 4, 5–6
development: imperative 206–8; paradigms 11–15; and sustainability 37; and underdevelopment 6–8
Development First approach 84, 230
development goals 174, 211, 212–13, 216; planning and development process 81–2, 83, 92, 99; UN Millennium 6, *7*, 54, 63, 81, 89, 137, 142, 206, 213, 228–9
development option 4, 16–17, 19, 28, 42, 99, 223–4, 231
development theories 11–15, 208–13
Dikhololo Resort, South Africa *41*
direct expenditure 181
displacement of communities 72, 130, 131–2
distribution of power 10, 129
distribution of wealth 6, 19, 85, 181
DMS (Destination Management System) 70
domestic tourism 1, 2, 19, 99, 158, 170–1, 221–2
Dominican Republic 26
Dubai 162
Duffy, R. 85, 103, 104–5, 106
dynamic packaging 24

e-commerce 70
Earth Summit 30, 137
East Asia Pacific 1–2, 24
Eastern Europe 59, 161–2, 198
EasyJet 59, 69
EC 175
eco-labelling 49, 51
EcoCertification Programme 51
ecological component of tourism 6
ecological-human communities 96, 121–2
ECOMOST project 53
economic issues: benefits 180–3; component of tourism 6; costs 184–5; dependency 184–5, 220; diversification 13, 170; fragility 9–10; globalization 63–70; impacts 180–5; integration 61, 63, 218; neoliberalism *12*, 13, 14–15, 35, 210, 218; significance of tourism 1, 2–3; sustainable development objectives 35; trading blocks 63
ecosystems 91, 179, 187, 189; capital 223; global 34; network model 96
ecotourism 103–7; *see also* green tourism
Ecumenical Coalition on Third World Tourism 134
educational opportunities 6, 195
educational tourism 159
effectors of demand 149, 167–8
ego-tourism 155
Egypt 24, 189
EIA (Environmental Impact Assessment) 97
El Jem, Tunisia *109*
embargo, US travel to Cuba 176
employment generation 183
empowerment 106, 116, 119, 130–1
enclave model of Third World tourism 72, 102
entrepreneurial activities 183, *184*
environmental issues 8, 14; awareness 163; concern 14, 32–3, 134, 148, 163, 164, 198, 230; conservation 164, 186; degradation 28, 33–4, 142, 171, 189, 198, 211; impacts 186–92; permanent restructuring 187; sustainable development objectives 35; values 156, 164, 169, 176, 178
environmentally appropriate tourism product 147–8, 164
equity 35, *36*, 48
escape motive 155
Estancia Santa Susana, Argentina *118*, *194*
Europe 22, 24, 59, 67, 69, 158–9; Eastern 59, 161–2, 198; Southern 176; Western 4, 217
European Blue Flag campaign 220
European Union 19, 63, 71

eviction *see* displacement of communities
evolution of development theories 11–15, 208–13
exotic destinations 1–2, 25, 73, 152, 156, 159–61, 198
exotic 'other' 58, 76
exploitation 26, 33, 144, 176
export 9
export led economies 13
externalities 185
extrinsic motivations 154–5

fair trade 135–7
fakalakalaka 138–9, 212–13
FAO (Food and Agriculture Organisation of the United Nations) 207
financial deregulation 59, 61, 64
financial incentives 112
First Choice 69
Fonatur, Mexico 86–8, 131–2, 134–5
Fordism 60, 61, 221
foreign capital 81
foreign exchange earnings 3, 9, 17, 170, 180–1
foreign investment 26, 59, 61, 91–2, 99, 106, 230–1
Formentera 97
Former Soviet Republics 16, 24
fragile environment 160
France 4–5, 101
free market 143, 229; economic neoliberalism *12*, 13, 14–15, 35, 210, 218; notion of comparative advantage 85, 217
Frequent Flyer Programmes 66
frontstage – backstage 118–19
full life paradigm 6
fundamental truths 44–5
futurity 35, *36*, 48

G-20 4–5
Galapagos Islands 75
Gambia, The 8, *46–7*, 49–50, 167, 180, 185, 195
gap-year holidays 141–2, 162–3
GATS 62
GATT 62
GDP 2, 4–5, 6, 8, 24, 182, *185*
GDS (global distribution systems) 70
gender and, community development 142–3
geographical production 73
Germany 4–5
global competition 13, 57, 63
global economic contribution of tourism 2–3
global growth and spread 158–9
global market 9, 22, 57, 61, 62, 63, 71, 72, 80, 211
global political economy 3, 4, 19, 49, 176, 206; *see also* International Political Economy (IPE)
global-local nexus 73–4, 224
globalization 218–19; cultural aspects of tourism and 74–8; economic aspects 63–70; political aspects of tourism and 70–4; process of 59–63; sustainable tourism and 54–5
GNP 6, 180, 208
Goa 97, 132–3
Goa Foundation 132
golf tourism 160
governance 45, 59, 62–3, 64, 71–2, 86
government *see* state
Grand Palace complex, Bangkok *107*
grass roots movements 54, 117, 176
Great Wall of China *21*
Greece 132, 160, 176
green consumer 163–4
Green Globe 21 51, 54, 220
green tourism 163–4; *see also* ecotourism
greening 192, 228
Grenada 216
growth impulses 13
growth poles 13, 27, 112, 217–18
growth of tourism 1–3, 17–19, 21–4
guest–host relationship 74–5
Guinea 63
Gulf War 21

Habitat for Humanity International 141
Hainan, China 108–9
Hall, C.M. 59, 86, 117, 120, 140; and Jenkins, J. 112; and Page, S. 95
Hard Rock Cafés 69
Havana, Cuba *160*
have-nots 129
health care 116, 224
health hazards 189
health scares 3, 21, 158
health/wellness tourism 160
heritage sites 19, *21*, 109–10, *186*, 192
heritage tourism 107–10
Hill Tribe people, Thailand 117
Hilton 91
holistic perspective 35, *36*, 47–9
Hong Kong 4, 24, 134, 222
hotels 15–16, 17, *18*, 53, 58, 88; Amir Palace Hotel, Tunisia *101*; vertical integration 67–9
human perceptions 215
human rights 208–9
Hungary 9

Ibiza 97
IBRD (international Bank for Reconstruction and Development) 62
ICT 70
ideologies 11, 33, 35, 82, 84–5, 86

IFTO (International Federation of Tour Operators) 53
IGCP (International Gorilla Conservation Programme) 135
Iguazu Falls, Argentina *186*
IMF (International Monetary Fund) 13, 62, 63, 89, 93
immigration 59
import propensity 184
import-substitution policies 11
income generation 181–2
India 4–5, 9, 19, 88, 142, 158, 159, 161, 197, 288; domestic tourism 170; Goa 97, 132–3
Indian Ocean 19, 161
Indonesia 4–5, *20*, 94–5; Bali 3, 94–5, 102, *184*, *197*, 217–18, *227*; Bangunkerto village *125*; Lombok *68*, *103*, *104*, *143*, *212*
industry initiatives 52–3
inequality 6, 9, 10, 55, 140, 210; equity principle 35, *36*, 48; reinforcement of 14, 58, 72, 122
inflation 185
informal sector 133
informal/indirect employment 17
infrastructure 19
integrative development, and alternative tourism 102–3
Inter-American Development Bank 99
InterContinental Hotel Group PLC 88
international agencies 89–91; funding 82, 138, 217, 218, 226
international arrivals/receipts 1–2, 5, 8, 17–19, 21, 24
international conflicts 21
international funding agencies 82, 138, 217, 218, 226
International Political Economy (IPE) 15, 73, 85–6, 229; *see also* global political economy
international tourism 1–2
Internet 60, 70, 219; booking 58, 59, 161; sites 24, 48, 141
intrinsic motivations 154
investment, foreign 26, 59, 61, 91–2, 99, 106, 230–1
Israel 4
Italy 4–5
ITP (International Tourism Partnership) 53
IUCN (International Union for the Conservation of Nature) 30, 33, 34, 135

Jafar's Four Platforms of Tourism 27
Jamaica 93, 101; Sandals *153*
Japan 4–5, 222
Japanese International Cooperation Agency 71
JGF (Jagrut Goenkaranchi Fouz) 132
Johannesburg 30, 122, 137, *139*
Jordan 24, 71
justice tourism 141

Kasongan pottery village, Indonesia 211
Kenya 77, 106, 117, 163, 189, 228
Kibera, Kenya 117
knowledge stage, Jafar's Four Platforms of Tourism 27
knowledge transfer 219
Knysna National Park 105
Kruger National Park 105, 106
Kuta Beach, Bali 102, *227*
Kyrgyzstan *16–17*

La Moka Ecolodge, Cuba *126*, *127*
labour 58, 63–4, *67*, 75, 116; casual 138; regulation 96, 97; women 142, 143
LAC (limits of acceptable change) 203
ladder of economic development 206, 211, 217
land contamination 189
landscapes 59–60, 73, 192
language 75, 76, 108, 116, 150, 197
Laos 24
Las Terrazas Tourist Complex, Cuba *126*, *127*
Latin America 11, 91
LDCs (least developed countries) 1–3, 8
Lea, J. 2, 71, 72, 176, 187, 189, 193
leakages 19, 83, 88, 101, 106, 184
life expectancy 8, 207
Lijiang, China 128
limits of participation 129–30
linkages 102; backward 19, 111, 183; and participation 110–12
literacy 8, 9
Llasa, Tibet 74
Local Agenda 21, 122–3, 128, 134
local community *see* community/communities
local elites 4, 13, 48, 97, 103, 111–12, 115, 117
Lombok, Indonesia *68*, *103*, *104*, *143*, *212*
low cost airlines/flights 58, 59, 61, 161
low-impact tourism *37–8*

McDonald's 69, 74, 77
McMichael, P. 62–3, 78, 210–11
Majorca 97
Make Poverty History campaign 6
Malaysia 24, 61, 94, 222
Maldives, The 8, 19, 163
Malta 95, 160, 202
manufacturing sector 9
marginality 77
marginalization 2–3, 81, 110, 123, 205
market liberalization 13, 59, 61, 62, 73
Marrakech, Morocco 59
Marriot 101

Masai Mara, Kenya 189
mass customization 61
mass tourism 25, 31, 58, 60–1; vs alternative tourism 39–40, 41; *see also* all-inclusive resorts
master plans *87*, *90–1*, 94, 98–9, 200, 223
Matmata, Tunisia *158*
media 59–60, 74, 76, 77
Mediterrranean 73, 89
Mercosur (South American trading bloc) 63
metropolitan/periphery political economic model 5
Mexico 4–5, 14–15; Cancún 26, 60, 61, *65*, *87*, 109, 184, 217–18; Fonatur 86–8, 131–2, 134–5
Mexico City 207
micro-credits 138–9, 230
micro-finance 138
Middle East 1–2, 24
MIGA (Multilateral Investment Guarantee Agency) 89–91
Minorica 97
models: of attitudes 120–1; of tourism development 97–8
modernization 11–13, 217–18
modernist-technocentric belief 33
Monastir, Tunisia *101*
Montego Bay, Jamaica, Jamaica *153*, 220
moralization of tourism 52, 163
Morocco 59
motivations 150–5, 215
multinational corporations 13, 55, 58, 82, 85, 86, 88, 94, 135, 176, 218, 225, 230; *see also* transnational corporations
multiplier effect 27, 102, 182, 210, 219, 225
Mundo Maya, Mexico 109
Myanmar (Burma) 1–2, 24, 176

NAFTA (North American Free Trade Agreement) 63
Namibia 40–1
Nassau, Bahamas *68*, *161*
national development strategies 2
National Ecotourism Strategy (Australia) 50–1
national parks 104–6, 135, *190*, 192
natural disasters 3, 21, 60, 184–5, 206
natural resources 9, 33, 43, 137, 191
nature reserves 192
neo-Fordism 61
Nepalese Himalaya, impacts of trekking in *190–1*
networks 48, 60, 63–4, 66, 69, 76; model of ecosystems 96
new tourism/tourist 25, 27–8
New Zealand 4, 141
NGOs 14–15, 53–4, 133–5
niche markets/products 60, 156, 157, 159–60, 161, 169
NNEs 8
no-frills flights *see* low cost airlines/flights
non-economic development indicators 8
North America *see* Canada; USA
nostalgia 61, 88
not-for-profit organizations 91
Nusa Dua, Bali 102, 217–18
Nusa Tenggara, Bali 94–5

Oceania 4
Ocho Rios 220
OECD 187, 191
oil crises 21
Okavango Delta, Botswana 102
Oneworld 66
Only One Earth 33
opportunity costs 185
organization of tourism 161–2
Our Common Future (Brundtland Report) 32, 34–5

Pacific: East Asia 1–2, 24; South 161
package holidays 102
participation/involvement 121–4, 128–30; limits to 129–30; and linkages 110–12
partnerships 27–8, 63, 66, 88–9, 94; community 111–12, 122, 129, 134; public-private sector 168; strategic alliances 58, 60, 66
passport to development 123–4
peace 15, 76, 198
perceptions 215
permanent environmental restructuring 187
Philippines, The 196
Phuket, Thailand 19
physical impacts of tourism 186–92
planning: changing approaches to 95–6; and regulating for development 112–13; scales in 93–5; types of 93
pleasure periphery 147, 152, 161
polar tourism 160
policy 92–3; import-substitution 11; pricing 168; state 81; tourism 31, 51, 82–3, 84, 86–7, 88
political aspects of tourism and, globalization 70–4
political dimension of tourism 6
political economy: global 3, 4, 19, 49, 176, 206; international (IPE) 15, 73, 85–6, 229; metropolitan/periphery model 5
political empowerment 130, 131
political structures 4, 10, 26, 82, 110–11, 130; and processes 179
Political Trilemma of the Global Economy 218
political upheaval 3

pollution: generation of 187–9; of poverty 38–9
Polynesia 24
population 9; impact on dynamics 191–2
Portugal 132, 160, 176
post-development school 5
post-Fordism 60, 61, 221
post-sovereign government 71
post-structuralism 210
postmodernism 221
Potter, R.B. 10, 78, 84, 219–20, 222
poverty 6, 38–9, 45; pro-poor tourism 54, 137–40; underdevelopment 6–8, 9–10
poverty reduction strategy papers 13, 63, 218–19
power: in communities 130–3; distribution of 10, 129; empowerment 106, 116, 119, 130–1; and values 84–6
powerlessness 211
Pretoria, South Africa *140*, *171*
pricing policies 168
Prince George Wharf, Bahamas *161*
private sector 88–9; public–private sector partnerships 168
privatization 13, 59, 66, 93, 110
pro-poor tourism 54, 137–40
product enhancement and differentiation 168
product space organisations 96
production 219–20
production costs 63–4, 131, 210, 218, 230
production systems *3*, 60
progress 6, 11–12, 33, 123, 200
prostitution 196
protected areas 106, 192
protectionism 62, 140
PRSP (Poverty Reduction Strategy Papers) 13, 63, 218–19
psychological empowerment 130–1
public-private sector partnerships 168
purchasing 161–2

Quebec Declaration on Ecotourism 105
Quintana Roo *87*

Rain Forest Aerial Tram 89–90
rationalism 59
receipts/arrivals 1–2, 5, 8, 17–19, 21, 24
regional identity 131
regulatory environment 96–7, 112–13
Reid, D. 3, 34–5, 55, 57, 72, 99, 111
religion 195–6
relocation of communities 72, 130, 131–2
residents' attitudes 120–1
resistance to tourism 132–3
resort development 99–102
responsible tourism 134, 147, 164, 169
restaurants 67–9
Richards, G. and Hall, D. 117, 119–20, 123, 136
Rio + 10 30
Rio de Janeiro 30, 122
Rogerson, C. 137, 139, 213
Roman Colesseum *109*
rural migration 195
rural regeneration 2
rural socieites 72
rural tourism 51, 112, 133
Russia 4–5; St Petersburg *52*
Ryanair 59

SADCC (Southern African Development Community) 63
safaris *40–1*, 102, 105, 189, 228
safety and security 10
St Lucia 97, 216
St Petersburg, Russia *52*
SALPs (Structural Adjustment Lending Programmes) 13
Samoa 1–2
San José 90–1
SAPs (Structural Adjustment Programmes) 13, 62–3
Saudi Arabia 4–5, 24
SAVE (Save Abandoned Villages and their Environment) 26
scale: and control of development 98; planning 93–5; and scope of tourism 1–2
Scheyvens, R. 130, 131, 141
Sea of Cortcz project *87*
search for a new paradigm *12*, 14–15
security 10, 15
segmentation of appropriate markets 168
self-regulation 52
Senegal 8
Sengiggi Beach, Lombok *104*, *212*
sex tourism 196
sexually transmitted diseases 76, 196, 198
Shanghai 111
Sharpley, R. 19, 31, 47, 108, 148, 150, 164, 185, 198
short-break holidays 161–3
short-haul tourism 59, 169
Sicily 160
Sierra Leone 63
Silent Spring 33
Silk Road *90*
Singapore 4, 24, 61, 108, 176, 222; Indonesia and Malaysian Growth Triangle 63, 73, 94
Sky Team 66
small firms 89, 128, 136
SMEs (small and medium-sized enterprises) 89, 136
social capital 14–15, 210

social component of tourism 6
social disempowerment 130
social disorder 211
social impacts 195–6
social inequality 72
social institutions 5, 11, 34, 94
social sustainable development objectives 34–6, *43*
socialist market economy 92–3
socio-cultural impacts 192–8
socio-environmental issues 26–7, 84
socio-political structures 10
soft tourism 27, 39, 105, 159
Solomon Islands 102–3
South Africa 4–5; Dikhololo Resort *41*; Johannesburg 30, 122, 137, *139*; Pretoria *140*, *171*; pro-poor tourism 139, *140*; Sun City Resort *100*
South Asia 133
South Korea 4–5
South Pacific 161
Southeast Asia 4
Southern Europe 176
souvenirs 19, 76–7, 108, 126–7, *171*, 183, *184*, 196, 197, 211, 222
space-product organizations 96
space-time compression 59
spaceship earth 33
Spain 132, 160, 176, 195
sport tourism 160
Sri Lanka 19, 196, *196*, 211
stakeholders 168
standard of living 9, *10*
standardization 24–5, 58, 59, 60, 64, 77, 159, 221
Start Alliance 66
state involvement 86–8
state policy 81
statist approach 11
strategic alliances 58, 60, 66
structural adjustment 13, 62–3
STZC (Sustainable Tourism Zone of the Caribbean) 91
Sun City Resort, South Africa *100*
sun-sea-sand holidays 24–5, 152, 159, 169
supra-territorial spaces 59
sustainability: difficulties in implementing ideals 227–8; imperative 213–15; indicators 201–3
sustainable development 39–41, 222–3; and community involvement 121–4; definitions 32–4; and impacts 200; principles 35, *36*, 47–9; principles and objectives 34–6, *43*; principles to practice 36–9; tourism as 47–9
sustainable tourism development 39–41; accreditation schemes 51; definition 42–4; and globalization 54–5; guidelines 50–1; in practice 49–50; pro-poor tourism 54, 137–40; voluntary sector/NGOs 53–4; weaknesses and challenges 44–7

take-off stage 12–13, 45
Tanzania 1–2, 77, 97
taxes 72, 97, 112, 182, 211, 225
technical definition of tourism 5
technological innovation 59
Telfer, D.J. 11, 73, 210; and Wall, G. 111, 183, 219
Teo, P. 73, 74, 78
terrorist attacks 3, 19
Thailand 19; Grand Palace Complex, Bangkok *107*; Hill Tribe people 117
Thomas Cook (JMC) 69
Thomason (TUI) 69
Tibet 74
Timothy, D. 96, 108, 129, 130; and Boyd, S. 109
TOI (Tour Operators' Initiative) 53
Tonga 138–9, 212–13
top-down economic model 11
top-down planning approach 81, 95, 205
Tortuguero, Costa Rica 141
tour operators 67–9, 168–9
tourism: definitions 5; and development 26–8, 81–3; drivers of 15–20; international trends and flows 21–4; supply 25–6; types 159–61
Tourism Concern 50, 52, 53–4, 91, 134, 183, 220
tourism demand 21, 24–5; changes in nature of 159–63; effectors of 149, 167–8; process 148–50; trends 157–9
tourism development: actors in 86–92; outcomes 110–12; strategies 98–9
tourism development dilemma 4; framework 216–29
Tourism First approach 84
tourism impacts: assessment framework 175–9; local community responses 198–200; physical 186–92; socio-cultural 192–8; and sustainable development 200
tourism management 200
tourism markets and products 162–3
tourism planning consultants 90–1
tourism policy 31, 51, 82–3, 84, 86–7, 88
tourism satellite account system 180
tourism stressor activities 187
Tourism for Tomorrow awards 50
tourist behaviour 92, 117, 120–1, 149–50, 157; green tourists 163, 164, *165*; influencing 165–9; *see also* motivations; values
tourist cultural enclaves 77
tourist culture 117, 215
tourist expenditure 19, 106, 185
tourist typologies 150–2

tourist-host encounters 193–8
tourists 92; impact of 178, 189; impacts on 198
townships, South Africa *139*, *140*
trade barriers 19–20
trade-offs 224–7
Tragedy of the Commons 33
trans-state processes 57
Transat A.T. 69
transitional economies *see* globalization
transnational corporations 65, 72, 73, 110, 132; *see also* multinational corporations
transport 19, 24, 25, 47, 48, 181, 216; global 57, 59, 64; local 117, *160*, 221; planning 86, 91, 93, 94, 97, 102; pollution 187–8
travel career ladder 148–9
Travel Foundation 26, 53
tsunami 3, 19
Tunisia *18*, 73; El Jem *109*; Matmata *158*; Monastir *101*
Turkey 4–5, 121, 170, 180

UAE (United Arab Emirates) 24
Uganda Wildlife Authority 135
UK 4–5
Uluru (Ayers Rock) 176
UN Conference on Trade and Development 2–3
UN Millenium Development Goals 6, *7*, 54, 63, 81, 89, 137, 142, 206, 213, 228–9
UNCHE 33
UNCTAD 1–3
underdevelopment 6–8, 9–10
UNDP human development index 6
unemployment 9, 183
UNEP 169, 189
UNEP/WTO 31–2, 50, 51, 105, 201
UNESCO 109
UNWTO *see* WTO (World Tourism Organization)
urban economic regeneration 2, 71
urban slum tours 117
USA 4–5, 22, 176

values: environmental 156, 164, 169, 176, 178; and power 84–6; and tourism consumption 155–6
Varadero, Cuba *18*, *100*, 101–2
VCP (Vietnamese Communist Party) 92–3
vertical integration 67–9
Vietnam 24, 92–3
village tourism 94–5, 102–3, 116, 136, 211, 214, 216, 224
voluntary sector 53–4
volunteer tourism 141–2

waste generation 187–9
water quality improvement schemes 192
WCED (World Commission on Environment and Development) 30, 32
wealth, distribution of 6, 19, 85, 181
West Africa 8
West Bengal 88
Western Europe 4, 217
Western imperialism 27, 84; *see also* cultural imperialism
Western Region Development Strategy 15–16
WestJet 59
wilderness areas 19, 105, 159, 160, 164
wildlife reserves 192
wine tourism 160
women, position of 142–3
World Bank 8, 13, 62–3, 71, 89, 93, 134, 135, 218
World Commission on Environment and Development (WCED) 30, 32
World Conservation Strategy 30, 34
World Conservation Union 30, 33
World Heritage Sites *21*, 109, *186*, 192
World Summit on Sustainable Development 30, 31, 122, 137
World Tourism Organization (WTO) 1, 2, 24, 42, 158–9, 192, 201, 221, 222; and tourism planning consultants 90–1; UNEP/WTO 31–2, 50, 51, 105, 201
WTTC (World Travel and Tourism Council) 2, 24, 27–8, 50
WWF (World Wildlife Fund) 50

xenophobia 198

Yemen *91*
Yogyakarta, Indonesia *20*

Zanzibar 97
Zimbabwe 49
zoning 94, 95, 106, 167